A FIELD GUIDE TO

THE FISHES OF KUCHING RIVERS

SARAWAK, MALAYSIAN BORNEO

A FIELD GUIDE TO
THE FISHES OF KUCHING RIVERS
SARAWAK, MALAYSIAN BORNEO

Katherine Atack

with photographs by

Ch'ien C. Lee

Natural History Publications (Borneo)

Kota Kinabalu

2006

Published by

Natural History Publications (Borneo) Sdn. Bhd. (216807-X)
A913, 9th Floor, Wisma Merdeka Phase 1
P.O. Box 15566
88864 Kota Kinabalu, Sabah, Malaysia
Tel: 088-233098 Fax: 088-240768
e-mail: info@nhpborneo.com
website: www.nhpborneo.com

in association with

DANIDA

Danish International Development Assistance (DANIDA) and the State Government of Sarawak

This project is a component of the Sarawak Government-Danida
Urban Environmental Management System (UEMS) Project
Natural Resources and Environment Board (NREB), Kuching, Sarawak

First published January 2006

A Field Guide to the Fishes of Kuching Rivers, Sarawak, Malaysian Borneo
by Katherine Atack
with photographs by Ch'ien C. Lee

ISBN 983-812-113-4

Half title page: Marble Goby (*Oxyeleotris marmorata*), locally known as *'Betutu'*.
Frontispiece: An 'archerfish' spitting at an insect; as the name suggests, these fish are famous for their shooting technique and marksmanship.
Title page: Mangrove Stonefish (*Leptosynanceia asteroblepa*).
Page v: Spiny Eel or Fire Eel (*Mastacembelus erythrotaenia*).

Printed in Malaysia

Contents

Foreword

The history of Kuching is closely linked to its river—the Sarawak River and its tributaries. In the past, the river provided access to the sea and thereby to the outside world. At the same time, it sheltered the city from the sea and the possible exposure from pirates and other intruders. Today, the recreational value of the river is priceless. Not only does the river serve as a means of transportation and recreation, it is the biotope for a rich and diverse aquatic flora and fauna. For fishes alone, more than 100 species have been documented in rivers within the city boundaries. As such, the health of this river ecosystem is important to all of us.

The Kuching waterfront is unique, and has been the identity and pride of Kuching city. However, the fast growth and development of the city pose a serious threat for the environmental quality of the river. Like many beautiful urban rivers in other parts of the world, the inefficient sewerage system may potentially turn Sarawak River into a stinking sewer. Therefore, it is a task for all of us to ensure that this will never happen to our beloved river. Not only is there a need to improve the existing sewage system, but also to change the behaviour of each of us as individuals.

This book presents one aspect of the present rich biological diversity: the fishes. The book documents the fishes identified in the Sarawak River during a fishing survey carried out by the Sarawak Government/DANIDA-Urban Environmental Management System Project 2002–2005. This is the first time we are provided with a comprehensive overview of the fishes that are living within our urban boundaries.

The guidebook provides an important tool for continuous monitoring of the biological life as an indicator for the health of Sarawak River and its tributaries. The field guide is also meant as a tool for everybody who wants to explore the biology of our rivers, either for recreation, as part of a school curriculum or for research. Equally, it is very valuable for those who just enjoy reading books about nature, admiring the impressive pictures and the descriptions of the fishes.

The value of this book actually goes far beyond the Sarawak River. The information we get from this book will be very useful for observations in any coastal estuarine environment in the entire region.

I am grateful to the author and photographer for their efforts to make this project a success, and also to the Danish International Development Assistance (DANIDA) for their support and inspiration.

I wish everybody a good time exploring the rich and diverse aquatic flora and fauna of the rivers around Kuching City.

YB DATUK MICHAEL MANYIN AK JAWONG
Minister for Environment and Public Health,
SARAWAK

Preface

This book is remarkable for several reasons. First of all, there are almost no books intended for a lay readership that offer a guide to fishes of Borneo. This is significant because no group of vertebrate animals has more impact on the lives of people. This is especially true for people living in a coastal city like Kuching because fishes constitute such an important element of the daily lives of coastal people, both as a component of diet and as a component of local commerce. Secondly, although Borneo has an extremely rich fish fauna, most studies and books on this fauna observe a kind of boundary, where studies on freshwater species are separate from studies on marine fishes. I know this boundary personally because I have worked on the fishes of Borneo, but always on the freshwater species. Yet along the Bornean coasts this boundary between fresh and saltwater is ignored by the fish fauna. Some fishes live more or less permanently in the brackish water zone affected by the tides and some marine fishes migrate into fresh (or at least brackish) water to breed. At least one Bornean species migrates to the sea from freshwater to breed. Thirdly, the author of this book, Ms Katherine Atack, has temporarily taken a detour from her usual research on corals and coral reefs, which put her in contact with the fauna of the waters of Kuching, to write a book on fishes. Most specialists resist pressure to look at other types of animals. Finally, since this book ignores the artificial boundary between salt and freshwaters, hopefully it will suggest to serious students of the fish fauna that there is much to gain by considering the flow of species and, therefore, of energy between these two kinds of environments.

This book was not written for the benefit of scientists, but rather for a lay audience—people with an interest in the world around them. Some terminology is inevitable in order to describe aspects of the basic anatomy of fishes. But that terminology is held to an essential minimum and contributes to the author's aim of making this fauna accessible to the reader. Hopefully, this book will add to the enjoyment and appreciation of living in one of the biologically richest environments in the world.

Robert F. Inger

PROFESSOR DR ROBERT F. INGER
Emeritus Curator of Amphibians and Reptiles,
Field Museum of Natural History,
CHICAGO

Introduction

This book is an attempt to record the fish life inhabiting the rivers of Kuching, greatly augmented by the beautiful photographs of Ch'ien C. Lee. I only hope that the text is as informative and, as such, complements the high standard of photography.

The devil on my shoulder told me to categorise this field guide according to colour, to appeal to the lay reader and to frustrate the expert. But experts need not despair; the angel on the other shoulder won, and the guide follows conventional categorisation. That same devil would also have had me tell the reader and researcher that certain species have been omitted because they have been declared extinct, or because they can easily be found on restaurant menus along Kuching's riverfront. Again the angel won; all fish that were found have been included, although you will indeed find many of them gracing your dish in the restaurants of Kuching!

References to fish can be found in many places, from the Bible to astrology and pre-Christian beliefs. The fish symbol often associated with the acrostic (word puzzle) ΙΧΘΥΣ (translated as Ichthys in English) written in the middle is commonly thought to have originated as a code symbol for early Christian communities. It is now known to have originated thousands of years before Christ existed (B.C.E.). It is thought to have originated in North India and be related directly to the sun; hidden knowledge and wisdom; Aries or Pisces; the astrological 'Dog Star' (Sothis or Sirius) and the Avatars or Neros (periods or cycles of time). Possibly, the earliest use of the fish as an icon was in the 'Cult of the Fish Mother' during the 6th millennium B.C.E. Since then, the fish has been used as an icon throughout the world: representing the Great Mother Kwan-Yin in China, the Goddess Kali in India, Isis the 'Great Fish of the Abyss' in Egypt, Delphos representing the fish and the womb in Greece, the Great Goddess Freya in Scandinavia, the Great Goddess Ephesus in the Middle East, Oannes the

(Opposite). Nipah palms growing along the banks of an estuarine river in Sarawak.

'Fish man' in pagan beliefs, and Ichthys was the son of the ancient sea goddess Atargatis.

Not only have fish been used as a symbol throughout mankind's history, but accounts and stories of sea monsters that could be heard as folklore, superstition, myth and legends abound and persist even in these pragmatic

Map of Borneo showing the location of Sarawak.

times. The Chinese have used fish to harness and direct *'Chi'*, or Nature's energy, the Romans kept eels as a sign of wealth, while many of us may remember exploring small rock pools along the shores at low tide during our childhood. Today, aquaria are everywhere, as miniature windows into the underwater world, and have become increasingly popular in the urbanised environment in which many of us live.

South-East Asia is one of the most diverse areas of the world for fauna and flora. Vast numbers of fish species can be found in an incredible array of shapes, sizes and colours.

The Island of Borneo is the third largest island in the world, situated on the equator in the South-East Asian archipelago. This rich, tropical paradise of complex ecosystems includes tropical rainforest, abundant coral reefs, and extensive mangrove habitats, which protect much of the coastline. Borneo is one of the most bio-diverse areas of the world, with many species being endemic either to the island itself or to the region. There are an estimated 15,000–16,000 species of plants, 360 birds and innumerable insects, with new species still regularly being discovered. In the island's rivers and lakes, there are approximately 400 species of freshwater fish, of which roughly 150 are endemic to Borneo. Estimates for marine fish alone vary from 1000 to 2000 species.

Borneo's terrain is generally low-lying, with over half the island below 150 m elevation. Tides, which in places may reach as high as 6 m, can penetrate inland as far as 100 km. The island has little seismic (volcanic) activity, other than areas of hot springs and mud volcanoes; however, bulging of igneous rock and folding of older sedimentary formations have combined to shape mountain ranges to over 4000 m elevation in the interior of the island.

The island is divided between three countries: the four Indonesian provinces of Kalimantan on the south and east of the central mountain ranges, the two East Malaysian states of Sabah and Sarawak to the west and north respectively, and the small independent sultanate of Brunei Darussalam on the north-west coast.

Sarawak covers an area of approximately 124,449 sq. km. Whilst most of the state consists of flat alluvial plains and lowland dipterocarp forest, the interior consists of a series of mountain ranges. Abundant freshwater rivers drain a number of habitats, such as hill and montane forest, lowland dipterocarp forest, peat swamp forest, freshwater swamp forest, and heath forest, with each supporting a correspondingly different composition of fish species.

Many of the coastal areas are lined with mangrove forests and muddy shores. The highly turbid coastal waters mean that coral growth is limited to the continental shelf off the coasts of Miri, Bintulu and some of the islands off Sematan. The inshore mangrove forests are important as fish breeding grounds, and thus essential for the commercial fisheries industry. This vital habitat is currently protected at Bako National Park, Similajau and Samunsan Wildlife Sanctuary.

To date, 254 freshwater fish species have been recorded from Sarawak, with new species being discovered regularly, and several currently awaiting publication. Only a few in-depth studies on the fish fauna have been carried out in Sarawak; these include the areas of Gunung Mulu National Park, Bau Limestone Region, and Baram. The Mulu survey, which was carried out in 1978, recorded 48 species, comprising 34 genera and 14 families. A survey in 2001 of the Rayu basin (Kubah National Park) discovered 27 species comprising 18 genera in 13 families. The recent Bau survey, in 2002, found 35 species comprising 25 genera and 15 families; six species were endemic to Borneo and two species new to science. There are two reports from the highlands: Watson and Balon (1984) found 57 species in the Baram area and Lee Nyanti *et al.* (1999) found 24 species (7 families comprising 19 genera) in the Bario area.

Sarawak's inland fisheries include a number of highly esteemed species. The Cyprinid (carp) family has many valuable members; the most valuable include those locally known as '*Empurau*' (from the Iban language), '*Semah*' (pronounced phonetically) and two species of *Puntius,* one with a red tail and one with a white tail. Species other than carp include the '*Tapah*' (*Wallago* sp.) and the herring locally known as '*Terubok*' (*Tenualosa* sp.). These fishes and many others are unfortunately threatened by activities such as over-fishing and habitat degradation, such as the draining of peat swamp forests for conversion into oil palm plantations, logging, prawn farms and coastal development.

Tor tambroides ('*Empurau*' from the Iban language) is a large herbivorous species of carp which lives in swift, clear, rocky streams in the upland forests. It was once found in most of Sarawak's major rivers such as the Rejang, Baram, Limbang and the Ai. However, its naturally small population makes it very sensitive towards uncontrolled fishing and loss of

The Sarawak River running past Kuching City's 'Waterfront" at sunset.

habitat, and consequently it has become scarce. It is a highly esteemed food fish which can reach a maximum of 100 cm in length (over 20 kg) and realises market prices over RM160 per kilogram. Attempts to reduce the consequences of pressure on the wild population by commercially farming this species via aquaculture unfortunately have so far not been viable due to problems with artificial breeding.

Tor douronensis (*'Semah'*), which has been given the title Sarawak's 'State Fish', is very similar to *T. tambroides* in both habitat and biology, but it has a slightly wider distribution and is smaller, with a maximum length of 40 cm. A highly popular market species, its population has also drastically declined due to overexploitation, pollution and general habitat loss.

There are two species of *Tenualosa* found in Sarawak, *T. toli* and *T. macrura.* The young of both species migrate far upriver, returning to the

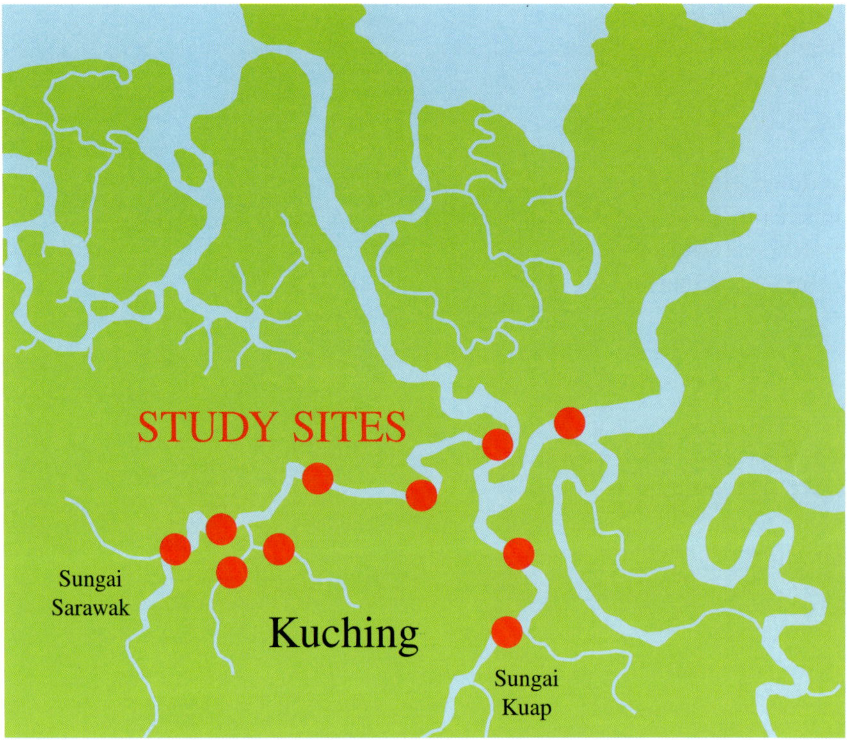

Map of Kuching river system, highlighting the fish survey areas.

estuaries only to spawn. The alteration of many of Sarawak's river systems has prevented the young fish from travelling upstream and has proven deleterious to the *Tenualosa* fishery throughout the state.

Kuching, the capital city of Sarawak, emerged as a small trading port and seat of the Brunei Government in the 1830s. It grew in importance during the reign of the 'White Rajahs', three generations of the Brooke family: James Brooke, Charles Brooke and Charles Vyner Brooke. The city resumed its importance when Charles Vyner Brooke ceded Sarawak to the British Crown in 1946, subsequent to the Japanese occupation during World War II. When Malaysia was formed in 1963, Kuching became the capital of the Malaysian State of Sarawak. Kuching is located close to the northwest coast on low-lying, generally flat marine clay soils and peat. The flat landscape is drained by a labyrinthine system of rivers meandering through this alluvial delta area to the coast, and is often flooded during prolonged heavy rains and king tides. The Sarawak River has been the heart of the town throughout this period. Today, Kuching is a rapidly expanding city with a population of approximately 500,000.

The rivers flowing through the city and the surrounding area are tidally influenced; many of them retain estuarine conditions some distance upstream. Almost all the major rivers have flat muddy or silty beds, turbid waters and slow currents. The fishes presented in this book were all collected from the stretches of the Sarawak, Kuap and Maong Rivers that run through Kuching city.

The gates of Kuching Barrage control the water level of the Sarawak River running through the Kuching area.

The Sarawak River is a relatively large river with two main tributaries: the Sarawak Kiri (left fork), which originates in the Penrissen highlands on the Indonesian border, and Sarawak Kanan (right fork); both these tributaries run through the Bau limestone region. The lower reaches are low lying and tidal with mangrove trees and Nipah palms (*Nypa fruticans*) along the undeveloped riverbanks.

In 1997, the Sarawak River became controlled with the launching of the Kuching Barrage, a tide gate structure located below the town just above the confluence with the Kuap River. At the same time, the outlet via the Santubong River was blocked off completely.

The Kuching Barrage forms part of the Sungai Sarawak Regulation Scheme. It was constructed for the purposes of maintaining a stable water level at the waterfront in Kuching proper, preventing tidal flooding in the low-lying areas of Kuching, and preventing saltwater contamination of the city's water supply. The Barrage comprises five radial gates, which may be closed against the incoming tides, and a ship lock, which allows ships to enter the upper Sarawak River.

Prior to the Barrage construction, the salinity levels changed daily, with a gradual decrease upriver as tidal influence declined. This pattern changed with the commencement of Barrage activities. When 'flushing' is carried out, a large influx of saltwater inundates the lower reaches of the river in a rapid ascent; once the gates are closed, this salinity slowly declines with the ingress of freshwater. This decline continues until the next time the gates are opened and the tide is allowed to enter again. When the control gates are closed, the waters of the Sarawak and Maong rivers upstream of the Barrage are relatively sluggish with no tidal influence.

This periodic flushing leads to a mixed population of freshwater and saltwater fish species upstream of the Barrage, which can change markedly depending on how frequently the gates are opened. This may limit the use of the river as a spawning route for species such as *Tenualosa* that migrate upstream to spawn and *Macrobrachium rosenbergii*, which migrate to the estuary to spawn.

The area of the Sarawak River that was surveyed stretches from Kampong Paroh, just upstream of the City, to the Barrage. Near the riverbank of Kampong Paroh there are relatively large fish cages which contain mostly aquaculture species of *Oreochromis*, commonly and locally known as *'Tilapia'*. The river bend at Kampong Paroh, at approximately 40 m, is the widest point of the river. There are small islands towards the centre of the river at this point with banks mostly covered with Nipah palms and mangrove trees. Large parts of the riverbanks from here down to the city proper are bare, with sporadic trees and palms. The banks along the city centre are reinforced with concrete, while a few large old trees line the north bank. The banks north of the city centre are least developed, with traditional Malay kampongs fronted with mangrove-like vegetation. This continues all the way to the Barrage. Downstream of the Barrage, the Kuap River joins the Sarawak River, which continues for a short distance before entering the South China Sea. The undeveloped riverbanks of this area are covered with Nipah palms and mangrove trees. The developed areas include industrial areas, commercial ports and docks.

The morphology of the Kuap River is similar to the Sarawak River, but the banks are much less developed, except for the industrial area of Sama Jaya. The riverbanks are densely covered with Nipah palms and mangroves. The area of the Kuap River surveyed is strongly influenced by the tides, with water levels changing between two and four metres with each tide.

The Maong River is one of the smaller rivers running through Kuching. It joins the Sarawak River approximately 500 m upstream of the city centre. It has two tributaries, Maong Kanan (Right) and Maong Kiri (Left). Both these tributaries run through densely urbanised areas of the city, and in many areas the riverbanks have been reinforced with concrete. When the Barrage gates are closed, the river flow slows and quickly becomes stagnant. The riverbanks and waterways commonly become overgrown with large aquatic plants and floating weeds such as water hyacinth (*Eichhornia* sp.).

The Project

This book has been prepared as a component of the Sarawak Government/DANIDA Urban Environmental Management Scheme (UEMS) 2002–2005. The UEMS Project is assisting the Sarawak Government in designing a monitoring programme for the quality of the rivers flowing through Kuching, the capital of Sarawak. The indicators to be developed for the monitoring programme include the presence and abundance of fish and prawn species.

To establish a baseline on the present composition of fish and prawn species, a fishing survey of the Kuching rivers was carried out in 2003–2004. Regular fishing using '*jaring*' (gill nets) and '*jala*' (cast nests) was used to ascertain population size of the fish and prawns caught. Fishing was carried out at 16 sites in the larger rivers of Kuching: Sarawak, Maong and Kuap rivers. This was repeated every two months for a year. Most of the species caught have been photographed, and a number of the photographs are included in this book. The species caught included members of freshwater, estuarine as well as marine families. The report on the findings of the fishing has been prepared for the Natural Resources and Environmental Board (NREB): 'Monitoring Abundance of Fish, Shrimps and Prawns in Rivers and Tributaries in Kuching (October, 2003–August, 2004)' (Atack, in press).

The fishing revealed distinct and statistically significant differences in species composition between the rivers. The differences corresponded with salinity levels, dissolved oxygen in the water and ammonium levels in the sediment. These two last parameters indicate the level of organic pollution.

- The Kuap River had the greatest diversity and abundance of fish and prawn species. It had the greatest number of marine species, most of which were temporary visitors that came in with the high tides.
- The Maong River had the lowest diversity and abundance of fish, and no prawns were collected. It contained only freshwater species.
- The Sarawak River is more complex, with a mix of freshwater and estuarine species. There were more freshwater species around the Kampong Paroh area, while more estuarine species were found between the waterfront and the Barrage.
- Approximately 76 species of fishes and prawns were found in the Sarawak River, 86 species in the Kuap River and only 10 species in the Maong River. In total, 114 species were found from the rivers of Kuching.

Based on this study, future monitoring will be determined using a selection of indicator species and abundance and diversity indices. This book was primarily prepared to facilitate the identification of the species for future periodical monitoring, but may of course be useful for anybody interested in the river fauna. The book presents all the species identified during the one-year survey period.

Ecology

Fish ecology is a large and complex topic due to the vast number of aquatic environments in which fish can be found from tropical to polar waters, freshwater to seawater, from high altitude lakes to deep-sea trenches. This chapter intends to provide a brief introduction to fish ecology and specifically to the ecology of the fishes described in this book. The ecological description of an individual species can range from general, for example 'freshwater', 'marine' or 'brackish', to highly specific, e.g., 'shallow pools in a highland rain-fed peat swamp forest', or 'within the coral structure of a tropical coral reef'. The two types of descriptions are often given the terms 'macro-habitat' and 'micro-habitat'. Examples of macro-habitat would include dipterocarp-forest river, or swamp-forest lake in freshwater environments and coral reef, mangrove and seagrass beds in the marine environment (the larger picture). Micro-habitat describes where in the macro-habitat the species is found, such as the riffles or pools in a montane-forest river, or within the coral branches or sandy patches of a coral reef (gives the specific details).

When referring to freshwater fishes, many are found only in a specific type of habitat, and some are referred to as endemic, i.e., found only at a precise location. A species may be endemic to a country, a specific area within a country or exclusive to an area on one side of a particular mountain. Marine species, on the other hand, usually have wider distributions, such as a coastline, or a specific sea or ocean, although water temperature is usually a restrictive factor limiting distribution to tropical, temperate or polar zones.

Salinity (level of salt concentration in water) is the critical factor that separates freshwater species from marine. Only a very small proportion of fish have become specialised enough to be able to live and thrive in a brackish-water environment where the salinity fluctuates between extremes on a regular basis. Brackish-water environments include estuaries (also known as estuarine environments) and mangrove waters found around the coastline where freshwater meets the sea. There are also fish that move between rivers and sea at a specific time in their reproductive cycle, namely either catadromous (fish that move from freshwater to the sea to spawn, including some species of goby) or anadromous (fish that move from the sea to freshwater to spawn, such as salmon and the local *Tenualosa* sp.).

In all types of watery habitats, many species adapt to living within a certain stratum of water. Fish that live on or near the bottom are described as 'benthic' or 'demersal' fish; those in the mid level or upper levels of a water column are described as 'pelagic'. In many cases, the appearance of the fish (morphology) and behaviour have evolved in accordance with adaptability to a given stratum.

Ecology of Freshwater Habitats

There are many different types of river ecosystems. Freshwater rivers can range from small, clear, pristine streams, which can run through elevated rocky terrain or primary rainforest. These tend to be shallow fast-flowing streams, with waterfalls, small pools, riffles and rapids. These gradually grow into larger, slower, streams and rivers and eventually into flooded forests and lakes at lower elevations. River water can be described as either 'soft' or 'hard', i.e., acidic or alkaline (limestone raises the pH to alkaline levels of more than 7) and depends on the geology of the river's terrain. The vegetation above the river can affect the organic load and temperature (shaded streams are cooler than rivers running through flat open land). The river bottom substrate can affect the species composition, for instance muddy, sandy and silty substrates will have more benthic feeding species than rivers where the substrate is rock or clay. All these factors add up to make unique habitats, which gradually merge one into another, and each habitat will have a naturally shifting balance in species composition as food supply and niche opportunities change with season and natural disturbances (drought, storms, trees falling, etc…). This naturally shifting balance is known a 'dynamic equilibrium'. When a foreign species is introduced, this balance maybe lost and devastation is likely to occur, with possible extinction of local species.

In major urban areas, the main rivers are often large, murky, sluggish and highly polluted waterways. Banks of smaller streams may be reinforced with concrete enclosing an extensive network of drains. These direct stormwater and sometimes sewage to the rivers and tributaries, and provide their own ecosystem for some very hardy species.

Many species have adapted to specific environments. For instance, small sucking loaches are adapted to the fast-flowing rocky streams, where they can cling to the rocks as they feed in the strong currents. Many of the gouramies live in areas that flood regularly and therefore are adapted to living in the stagnant and low-oxygenated waters of stranded pools. Some gouramies and the air-breathing catfish can even bury themselves in the mud

Clear freshwater swamp typically found in Sarawak's rainforests.

when these pools evaporate and survive for weeks until the area is flooded again.

The coastal plains of Sarawak support extensive areas of rain-fed peat swamp forest. Peat soil has more than 65% organic content, and is usually at least 50 cm thick, but can be as much as 20 m deep. All incoming water is from rainfall, low in mineral nutrients, and the drainage water is acidic with a pH of 4 or lower. These waters are usually tea-coloured and described as 'black water'. Towards the centre of the swamp the streams become more acidic, resulting in lower species abundance and diversity. However, some species do adapt to this harsh environment; such species are often restricted to peat swamp habitats. For instance, the rare Arowana (*Scleropages formosus*) is found in the deep pools of peat swamp rivers. Many fish have evolved bright colouring to enable them to stand out in these dark waters and hence have become favourites of many ornamental fish collectors.

The river-fed freshwater swamp forests are found in the low-lying plains, where the rivers are broad, winding ribbons of brown water. In Sarawak, these are usually low-lying dipterocarp forests that become flooded with freshwater from inland lakes. The pH is usually above 6 as this water is rich in minerals, and therefore has much greater productivity than the rain-fed peat swamps, hence a much greater species diversity and abundance. Examples of this are the basins of the Rejang and Baram rivers. Generally, the riparian forests in Sarawak are seasonal swamp forests, i.e. they flood with the heavier rains from November to March.

Sarawak also has several extensive limestone formations, the best known being Mulu and Bau. The rivers that run through limestone are characterised by hard (high calcium content) alkaline water (with a pH higher than 7). The left and right tributaries (Sarawak Kiri and Sarawak Kanan) of the Sarawak River both flow through the Bau Limestone Region, and this contributes to the diversity of fish species.

Ecology of Estuarine Habitats

Estuarine or brackish-water habitats which include estuaries, mangroves and tidally influenced swamp forest are complex, highly variable and dynamic ecosystems with both freshwater and saltwater influences. The incoming tide brings an increase in salinity and water levels,

(Opposite). A mangrove swamp at low tide.

which later decrease with the outgoing tide. The rivers bring a constant flow of freshwater and nutrients, which not only lower the salinity, but also make the estuarine environment potentially much more productive than the surrounding sea. The flora and fauna in these habitats must be able to tolerate these widely fluctuating conditions.

An estuary occurs where a wide river enters into the sea, often in a sheltered bay. In Sarawak, estuaries are fringed mostly with Nipah palms. Small trees and shrubs of Avicenniaceae (*Avicennia* spp.) can be found along the riverbanks as far inland as the high tide's saltwater influx reaches.

The term 'mangrove forest' is the collective name given for the types of tree vegetation found in the brackish water environment. These are trees which have adapted to the high levels of salt and the fluctuating water levels such as those found in the delta of the Sarawak River. Mangroves are important physically, as a barrier between the sea and the coastal land,

Nipah palms and mangrove species line the banks of tidally influenced rivers in Sarawak.

protecting against erosion and the large waves from storms and tsunamis. Both estuaries and mangroves are biologically important as nursing grounds for many species of juvenile fish. Many commercial fish like grouper, croaker, snapper, horse mackerel, grey mullet and milkfish are typical migrants into mangrove waters, spending certain periods of their life cycle in this ecosystem. Mangrove swamps provide a relatively safe haven and a food-rich environment, ideal for nursery and spawning requirements. The maze of stilt and prop roots provide ideal conditions for egg-laying, and the tidal waters are enriched with oxygen. Eggs, fry, and juvenile fish are relatively safe from predators in these generally turbid waters. The onset of rains is a trigger for many fish species to migrate into mangrove waters. During this period, salinity is reduced and nutrients such as macro-algae, zooplankton and micro-benthic organisms are abundant, providing a rich and bountiful food supply.

There are many species of animals associated with estuarine environments and the adjoining rivers. The Estuarine Crocodile, *Crocodylus porosus*, favours the muddy and sandy banks of mangroves and estuaries, and can ascend far upriver. The strange-looking '*Ruang*' is a single-clawed crab shaped like a large prawn which burrows deep into the soft soils of the flooded mangrove, creating miniature volcano-like mounds with excavation debris. Mammals such as the Smooth Otter, *Lutrogale perspicillata*, and the Asian Short-clawed Otter, *Amblonyx cinerea*, and estuarine dolphins such as the Irrawaddy Dolphin, *Orcaella brevirostris*, are also common inhabitants of many sheltered estuarine and mangrove bays. Another very strange inhabitant is the king crab or horseshoe crab, *Carcinoscorpius rotundicauda*, which is not actually a crab but an ancient marine relative of the spider and scorpion with fossils found over 200 million years ago, a veritable living fossil! Numerous species of fish have adapted to this dynamic environment such as the highly territorial mudskippers, the mangrove stonefish, and the spitting archerfish.

Ecology of Marine Habitats

The number of different marine ecosystems is incredibly varied, ranging from coral reefs to rocky reefs, continental shelves to deep-sea trenches, tropical, temperate and polar seas, up-welling and down-welling currents, and thermal vents. Tropical marine ecosystems in Borneo

ROGER MUNNS/SCUBAZOO.COM

Algal beds thrive in clear shallow waters with lots of light.

(and Sarawak in particular) include coral reefs, seagrass beds, the continental sea floor, and pelagic waters over both the continental shelf (coastal) and the open ocean.

Coral reefs develop in warm, shallow, clear, and low-nutrient waters. They can be defined by their structure with fringing reefs developing adjacent to a coastline, while atoll reefs (circular reefs that have risen from deep-sea platforms such as submerged volcanic sea-mounts) are found far off any land mass. They are highly complex structures with millions of tiny coral animals building up a vast calcareous skeleton through generation upon generation. This vast matrix of interconnecting structures allows a wide variety of niches for all sorts of marine life, from tiny gobies and damsel fishes living in or around the small spaces within this coral framework to large fish grazing on the coral or hunting the smaller fish. Fish which can hide in fantastic places include tiny gobies and perchlets that live in the anus of sea cucumbers, on the whip fronds of corals, the arms of feather-stars and starfish, and of course the famous 'clownfish' that live among the arms of anemones. Other well-known reef fish include the larger, colourful coral-grazing parrotfish and wrasse, butterflyfish, surgeonfish, pufferfish and porcupinefish, triggerfish, trevallies and snappers. The assortment of fish species that inhabit this ecosystem is stunning.

Seagrass beds are, as the name suggests, beds of sandy substrate covered with grass-like algae plants in shallow, well-lit waters. Common species include *Halodule*, *Halophila*, *Enhalus* and *Thalassia*. These seagrasses stabilise offshore sand reservoirs, and are potentially a source of commercially valuable algae. Like mangroves, seagrass beds play a vital role as nursery to many species of juvenile fishes, such as rabbitfish, goatfish and mullets. They are also famous as habitats for the sea mermaid (a large marine mammal) known as the Dugong (*Dugong dugon*).

The sea floor of the continental shelf (whether, sandy, muddy or rocky), is home to a variety of benthic fish species, including tonguefish, soles and flounders. Goatfish, catfish and the predatory scorpionfish can bury

Tropical coral reefs abound with fish of all shapes and sizes.

themselves in the soft substrate or hide amongst the rocks and seaweed or actually camouflage themselves by mimicking their surroundings.

The open ocean has its own composition of fish species such as the smaller schooling plantivorous species including anchovies and sardines. These feed on the phytoplankton and zooplankton found in the upper surface waters. These smaller species are prey to the larger faster predatory trevallies, tunas and marlins.

Many of these species occupy several habitats throughout their life cycle. Juveniles grow and mature in the more protective environments of estuaries, mangroves and seagrass beds before migrating to open waters and reefs as adults. Species that are usually associated with open oceans occasionally visit coral reefs or estuaries where food is more abundant. Therefore estuarine habitats, such as found in the vicinity of Kuching, have a wide variety of fish species, ranging from permanent residents to temporary and even long-term visitors from both upstream freshwaters and the sea, that come to feed, breed or grow to adulthood.

JASON ISLEY/SCUBAZOO.COM

Large predatory Trevallies cruise the open oceans and outer waters of reefs.

Biology

Fish are cold-blooded vertebrates (animals with backbones). They are the largest class of vertebrates to be found in the aquatic environment. (Other classes of aquatic vertebrates include amphibians, reptiles, birds and mammals.) Fish are unique in that they are the only class of vertebrate animals to have gills as adults. To date, more than 20,000 species have been described worldwide.

Apart from the ecological environment, biology and behaviour are factors that characterise an individual fish species. Physiology, morphology, feeding, reproduction, the degree of parental care, and movement are all factors that combine to give each species its unique characteristics and its own place, or niche, in the environment in which it dwells.

Most fish are classified as either marine or freshwater and cannot survive in both habitats. There are, however, a number of species that either visit or have evolved to live in brackish waters and are therefore able to survive in both fresh and marine conditions. These include the majority of species found in the waterways around Kuching.

Fish Physiology

Fish are the only animals to actually breathe in water throughout their life cycle. As there is less oxygen available in water (<5%) compared to air (21%), fish have developed a specialised mechanism for oxygen extraction, namely the gills. The gill structure is highly advanced in the bony fishes. The gill membrane has a dense concentration of blood vessels through which blood flows in the opposite direction to the flow of the water. In many cases, numerous gill flaps, which can arch out, increase the surface area available for absorbing the dissolved oxygen from the passing water.

In many habitats where oxygen levels are even lower, for instance stagnant water bodies such as drainage ditches, canals, rice fields or slow-moving stagnant streams and swamps, some fish have evolved more advanced breathing methods using atmospheric air. These advanced methods may either be: 1) modification of the inner mouth, as found in Channidae (Snakeheads), 2) the presence of a specialised structure above the gills, often referred to as a labyrinthine organ (due to its maze-like appearance), as found in the Osphronemidae (Gouramies) and some species of catfish such as the genus *Clarias* (air-breathing catfish), 3) the modification of the digestive tract to enable the trapping of air as used by Synbranchidae (swamp eels), or 4) cutaneous respiration (breathing through the skin) as used by many of the mudskippers of the family Gobiidae.

A close look at the specialised skin of a mudskipper, used to breathe above water.

Almost all teleosts (group containing most of the bony fishes; this is the highest super-order of the ray-finned bony fishes) have a swim bladder, a sac where gases can be accumulated and released from the body. The swim bladder is usually found above the gut, and its function is to maintain balance as well as regulate the depth at which the fish swims.

Scales are made of a chitinous material similar to fingernails; their main function is to protect the fish and to help in reducing water uptake that occurs through osmosis. Not all fish have scales; for instance, eels and catfish go without. In some fishes, the scales have developed into armour, such as in the box and porcupine fishes. In almost all fish, the scales and skin contain

An example of chromatophores in the skin of a pipefish producing different colours and patterns.

chromatophores, colour pigmented cells that have a radial 'root-like' system where the coloured pigment can be dispersed throughout the cell or contracted into a point in the cell centre. When the pigment is dispersed, the fish appears darker, and when the pigment is concentrated, it appears lighter. There are many types of chromatophores and various patterns in which they can be arranged, hence the diversity seen in the patterns and colours displayed even within the same species, between males and females, juveniles and adults. The colouration displayed can be affected by environment, health, stress, fear and / or the stages of the reproductive cycle.

Fish Morphology

Fish swim with the use of fins; forward motion is driven with the caudal or tail fin, the dorsal and ventral fins help maintain balance, while the head and pectoral fins are used for control. When the dorsal and anal fins become fused with the caudal (tail) fin, the fish will adopt a snake-like motion. The small fleshy fin behind the dorsal fin (which has no rays) is called the adipose fin, and most commonly seen in the catfish. The various shapes and sizes in which fins are found can help determine the life style of the fish. An illustration of the tail fins and their descriptive names can be seen on page 24.

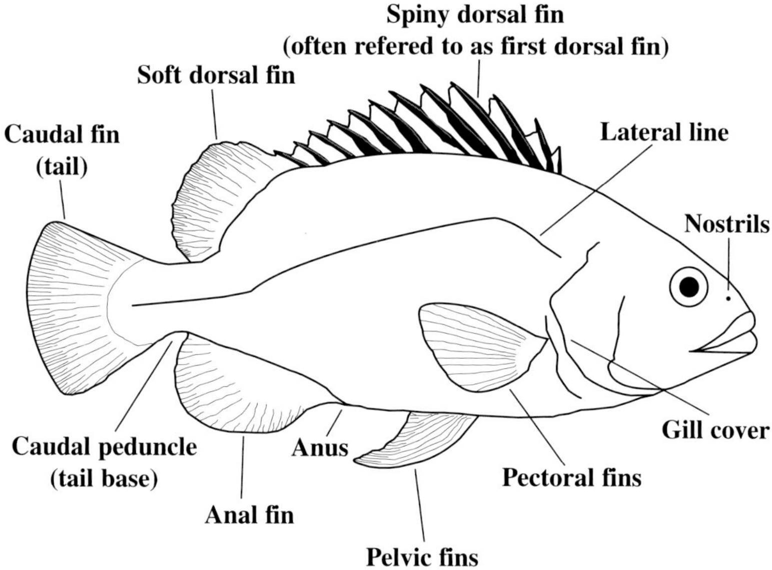

A general outline of the characteristics of a fish.

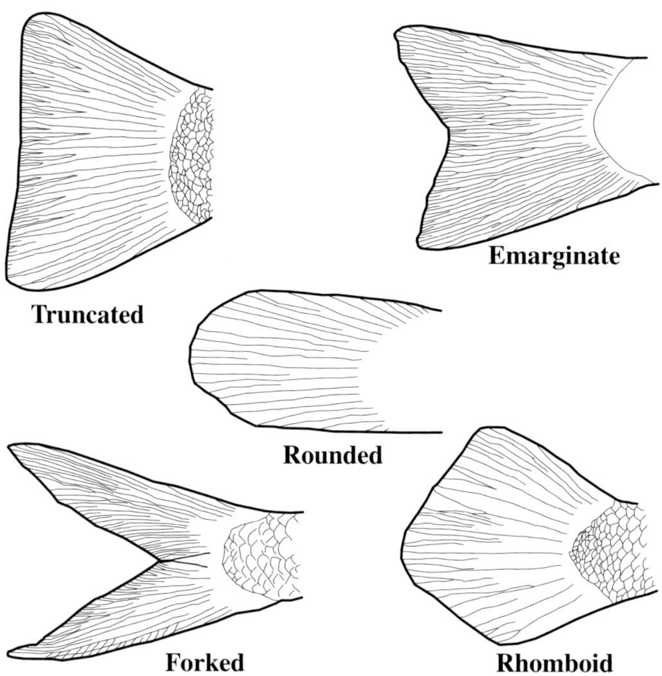

The different tail (caudal fin) shapes found in fishes.

Benthic fishes (that live on the bottom of the river or sea) can often be identified by their exceptionally flattened bodies, from extremes such as Cynoglossidae (flatfish), to less extreme body forms, where only the abdomen is flattened, such as many of the catfish, stonefish and toadfish. On the other hand, pelagic fish (fish that swim in the mid- and upper-waters), such as tuna, are built more for speed and manoeuvrability, having slender, ventrally-compressed streamlined bodies with long forked tails and long pectoral fins.

Reproduction

There are two forms of reproduction, internal and external fertilisation. Internal fertilisation is where the eggs are fertilised inside and remain inside the female until they hatch, when the young are ready to strike out (survive) on their own. External fertilisation is where the eggs are released and then fertilised into the surrounding water. Although this sounds simple enough, there is a vast variation in courtship, mating, spawning and parental care, from aggregate spawning with no parental care to courtship, nest building and parental care of both eggs and juveniles (fry).

In the marine environment, the most common form of reproduction is external free spawning, where eggs and young fry are dispersed by currents. Females and males, either in pairs or groups, make rapid dashes to the surface where they release their respective eggs and sperm. The eggs are small and produced in such numbers that, despite a high mortality rate, chances are high that a small percentage will survive to reach maturity. This behaviour is usually regulated with the time of day and phase of the moon.

In other species, the female lays her eggs on the benthic substrate where the male will then fertilise them. Parental care is holistic, including some degree of selection and early preparation of the substrate prior to laying of the eggs to provide protection and a clean environment. Some species build a nest or bury their eggs, and these are often guarded from predators and kept clean of dirt and algae. Some species only guard their eggs until they hatch, while others will guard the young fry for a certain period to increase the chance of survival. In some species, the male will pick up the eggs after fertilisation and keep them in his mouth until they hatch (often not eating throughout this period). Even after hatching, the male will keep an eye of the young fry and use his mouth as a sanctuary until they are big enough to defend themselves. Generally, the more effort parents invest in the next generation, the fewer and larger eggs are produced.

Many freshwater species have extraordinary reproductive behaviour and therefore have become a favourite of oriental fish collectors. For example, many members of the Cichlid family make nests and vigorously guard them. The nest may be an elaborate structure made of the surrounding weeds or from air bubbles trapped by the fish's saliva. These nests can be aggressively defended, for instance the Channidae (Snakeheads) will even attack humans if they approach too close.

Ovo-viviparity is the term used where species have adopted internal fertilisation; the female keeps the eggs inside until they hatch. Males of these species have therefore evolved modified anal or ventral fins that allow sperm transfer. Halfbeaks are a good example of this, and the Syngnathiformes (seahorses and pipefish) has taken it a step further; the female transfers the eggs to the male who keeps the eggs and fry in a specialised pouch.

Behaviour

Most fish behaviour largely depends on habitat, diet and reproductive methods. The habitat is important in terms of areas to hide, shelter and provide food. Whether the fish is predatory or not will determine when and where it eats, and whether it focuses on how to catch prey or how to avoid becoming prey. Fish have fused jaws and cannot chew; they either bite off chunks or swallow their prey whole.

With respect to diet, fish are herbivorous, carnivorous or omnivorous. Herbivores eat aquatic plants, small or large algae such as seaweeds and sea-grasses, fruits and vegetation that have dropped into the water. Carnivores feed on other animals such as benthic organisms, micro- or macro-invertebrates, crustacea, other fishes and even animals that live above water such as insects, snakes and lizards, amphibians and, occasionally, birds. Some carnivorous fishes prey exclusively on other fish and these can be described as 'piscivores'. Many species of fish that live in the surface waters of the sea feed on plankton (microscopic plants and animals found in the surface waters of the oceans where light is plentiful; both zooplankton (animals) and phytoplankton (plants)) and can be described as 'planktivores'.

Life for fish in the wild is a continuous battle for survival, with a fine line between having to eat and being eaten. Generally, the smaller the fish the more vulnerable it is to predators. Therefore, many species have developed ways to ensure that enough survive for the continuation of the species. Camouflage, used by both predators and potential prey, includes the use of cryptic colouration to enable the fish to blend into the surrounding environment; stonefish are among the best examples, looking so like the rocky substrate and remaining so still that divers will not see them even when nose to nose. The leafy sea dragon is a good example of avoiding predators: the leafy appendages on the body of this fish break up its outline and resemble the seaweed in which it dwells. Mimicry is commonly used to deter and confuse predators by adopting the colour and behaviour of other poisonous or dangerous species. Knives, spines and poisons are used by many species to deter predators, such as the scalpel-sharpness of the surgeon and unicornfish, spines of the porcupinefish and lionfish together with poisons in the spines of catfish and stingrays. Toxic flesh is also a good deterrent used by many species of pufferfish. Schooling in large numbers reduces individual chances of being eaten and also can confuse predators with the constant and coordinated movement and shifting of so many potential prey. Many predators use a combination of camouflage and ambush and / or speed to pursue and catch prey. The teeth of predators are usually sharp and pointed for gripping the smooth slippery sides of their prey.

Methods of reproduction affect the behaviour of different fish species. Fish that make nests and defend the eggs tend to be more territorial than species that spawn in the open and leave the currents to disperse their offspring. Mating behaviour is also important; for instance, there is very little courting in species that mass spawn, as both the eggs and sperm are released and there is very little control over which sperm fertilises which egg. In nest-building species where the female lays the eggs in a cluster, there is usually some courtship and / or battle to win the female, with the strongest or the brightest coloured male usually winning the right to fertilise the eggs.

Fair&Lovely

Fish and People

Fish have always been an important source of protein in the diet of human populations who dwell in coastal areas or alongside rivers and streams. In some cases, especially remote islands, fish are the only source of protein. Some fish or fish produce constitute delicacies that have gained worldwide popularity, such as salmon, tuna and caviar (the eggs of the sturgeon fish), while advances in technology have brought to light new delicacies such as those produced from the rare deep-sea Orange Roughy, *Hoplostethus atlanticus*, found around the deep-sea mounts off New Zealand (750–1200 m).

Before the advent of freezers, fish produce had to be either cooked fresh or preserved. Preservation and flavouring are still mostly carried out by salting, drying and / or smoking. Many delicacies utilise extremely fresh raw fish, for instance Japanese sushi and sashimi. One of the more unusual preparations is the use of non-fresh or fermented fish, esteemed locally as '*rusip*' and in Scandinavian countries as '*surstromming*'. Many of the fish found in the rivers of Kuching are esteemed for their flesh, and are often prepared by cooking as well as being salted and made into fishballs or stews. Traditional recipes for local fish varieties, even though fascinating, are beyond the scope of this book.

Markets and Fisheries

Local markets throughout Malaysia are important outlets for the small subsistence fishermen who sell their excess catch. The larger coastal fish markets are important for the big commercial fishing fleets, which include vessels from Japan, China and even Russia. Inland markets sell freshwater species and sometimes preserved marine fish brought in from

(Opposite). A distinctive sight along the Kuching Waterfront: sampans ferry people across the river.

(Above). The selling and buying of fish at the Kuching market.
(Opposite). Freshwater fish species sold at the local market in Kuching.

coastal areas. Coastal fish markets are usually much bigger and busier with fish generally caught on a much larger scale than can be caught from rivers; for example, schools of mackerel, sardines, and jacks can number hundreds to thousands of individuals.

Approximately 300 marine species of fish are caught throughout Malaysia, but only 100 of these are actually sold in the markets. The most esteemed groups of marine fish in Malaysia are snapper, bass, jacks, mackerels, tuna, pomfrets, the Clupeiformes (herrings, sardines and anchovies), sharks and rays.

Fishing Methods

Methods devised for catching fish throughout the world's small subsistence fishing communities are highly diverse: from fish traps and cages to nets of various shapes and sizes, cast netting, hook-and-line, spears and herding. The larger commercial fisheries have more standard fishing methods, with slight variations such as encircling nets (purse seine, haul seine, trawl seine), entangling nets (gill and trammel), and long lines. The fishing industry in Sarawak earns over 100 million ringgit annually and employs thousands of people; the most important commercial fishing methods are trawling, gill nets and purse seine nets. Local small-scale fishermen mostly use gill nets, cast nets, traps and hook-and-line.

Cast nets, locally known as '*jala*', are the most common method of catching prawns and small fish in the shallower waters of rivers and estuaries. A difficult technique that requires practice!

Gill nets, locally called '*jaring*', are available in different lengths and mesh sizes. Some nets have three layers with both large and small mesh sizes; these tend to be used in larger more estuarine rivers. The basic single-layer one-size mesh net is usually used in freshwater rivers. A net's placement is important; for instance, those placed across a river can only be used when the currents are favourable (not too strong). They may also be placed across the mouths of small tributaries, and sometimes parallel along the riverbank, where the catch then becomes trapped as the tide falls.

Cast netting, locally called '*jala*', is commonly used for catching smaller fish and, more importantly, prawns. It is a method generally used in shallower waters and it requires skilled technique. The net is circular with small lead weights sewn around the circumference. The net is thrown such that it spreads out and lands flat on the surface where the weighted edges sink faster and trap anything in that volume of water.

Hook-and-line is also very popular among sports fishermen; it is carried out both from small boats and from riverbanks. Certain bridges are choice spots and are often crowded with fishing lines hanging over the rails.

Two women using hook-and-line from a boat on the Sarawak river. Hook-and-line can also be used from river banks and bridges.

The Aquarium Trade

Many people believe that keeping fish in aquaria brings about harmony in their environment. Carp have been popular as ornamental fish for thousands of years, and are regarded as a sign of wealth, health and a bringer of good fortune and prosperity in Chinese households. During the last century, fish have been kept for ornamental display. Easy-to-care-for and very colourful goldfish were prizes to be won at carnivals and festivals and kept in simple glass bowls. As the aquarium trade developed, equipment became more elaborate, allowing the keeping and rearing of tropical freshwater species.

Keeping tropical fish has become a passion in many countries, and people have developed specific interests in certain fish families. For instance, Killifish clubs are now found in a number of countries worldwide; the aim is to promote the propagation, study and conservation of the African Killifish (of the order Cyprinodontiformes). This is a small, very brightly coloured fish which can be bred easily in captivity. Other favourite oriental freshwater species are Arowanas (Osteoglossidae), Cichlids, and Rasboras.

A display of aquaria, each promoting a glimpse into a fish's world. Many people find the presence of a well-kept aquarium calming.

Advances in technology and biological knowledge have led to an increase in the number and complexity of domestic marine aquaria, especially small replicas of coral reefs. Clownfish are a favourite, as are other members of the damselfish family as well as butterflyfish, brightly coloured gobies and surgeonfish. However, in many cases the environmental pressure caused by indiscriminate collection of wild specimens has pushed many a species to the brink of extinction.

Conservation

Since it is difficult to target conservation efforts at a single species of fish, protection is usually most effective when it is focused on habitat preservation. This can be difficult as a single river passes through many different types of environment throughout which a single species may range, so saving a particular part of a river is not always enough. The artificial culture and breeding of marketable and ornamental species are sometimes a viable method of reducing pressure on wild populations.

Pollution of rivers from industries, households and untreated sewage discharge has affected many fish populations. Habitat development and sedimentation caused by logging and land development are also factors affecting declining fish diversity. The loss of tree cover and the increasing sediment load in rivers result in warmer waters, which retain less oxygen and become a poorer environment for many species.

The role of fishes in our aquatic ecosystems is poorly understood even today. More research on the existing fish and niche requirements is needed before conservation plans can accomplish their purpose.

Methods of reducing river habitat degradation may include reduction of chemical and biological pollution by tighter regulations on industries and better sewage treatment facilities. In areas of logging, development or agriculture, the implementation of procedures such as river buffer zones (maintaining a certain area of original riverbank vegetation), and use of more advanced, ecologically friendlier methods in clearing, construction and processing have been proven to reduce sedimentation and help retain biodiversity.

Classification and Diversity of Fishes

There are some good reference books that can be used to identify the freshwater fishes of Borneo. These include *The Freshwater Fishes of North Borneo* by Inger and Chin (1990), *The Freshwater Fishes of Western Borneo (Kalimantan Barat, Indonesia)* by Roberts (1989) and *Freshwater Fishes of Western Indonesia and Sulawesi* by Kottelat *et al.* (1993). These books mostly cover the inland freshwater fishes and are not so relevant for the lower coastal reaches of rivers. Kottelat's book is the only one that includes clear photographs of fresh specimens. The marine identification books *Marine Fishes and Fisheries of Malaysia and Neighbouring Countries* by Abu Khair Mohammad Mohsin and Mohd. Azmi Ambak (1996) and *Marine Fishes of South-East Asia* by Gerry Allen (2000) are also very useful but, again, they do not cover all the fishes in the estuarine and lower rivers.

This book is designed to be a guide for the field identification of those fish species found in the estuarine habitats and lower reaches of the rivers in and around Kuching and, to some extent, other parts of Sarawak as well. It provides a brief description of each of the taxonomic orders and families that have been found in the rivers of Kuching. For the majority of species covered there is a clear photograph of a live or fresh specimen.

The scientific name is given in italics; the common name (English) is given, followed by the vernacular or local name (Malay or Iban), which is written in italics and parentheses. The physical description includes the standard length (this is the length from the snout to the base of the tail), the morphology (body shape, fin shape, spine and ray counts) and the colour. The description is of live or fresh fish specimens in all cases. Only the external characteristics are included as these are the important features for field

---

(Opposite). Face to face with the stunningly ugly Mangrove Stonefish (*Leptosynanceia asteroblepa*).

identification. The other books listed above have greater scientific and taxonomic content that can provide full external and internal measurements.

The description of distribution includes where in the rivers of Kuching, as well as globally, the species is found. A note is also given if the species is alien (introduced) together with its mode of introduction. Information on the habitat preferences and biology is given where known. In cases where similar species may possibly be present in the region, even though they have not been recorded or photographed, notes of the difference in characteristics have been included.

Taxonomic classification of fishes is as follows: Kingdom (animal), Phylum (Chordata), Class (Actinopterygii (ray-finned fishes) or Elasmobranchii (sharks and rays)) Order, Family, Genus and Species. The names of orders always end in '-formes' (shape in Latin) such as 'Anguilliformes' (meaning 'eel-shaped fishes'). The names of families always end in '-idae', such as Anguillidae. A family usually contains one or more genera and each genus may have one or more species. For example, in Borneo the genus *Rasbora* belongs to the family Cyprinidae, and there are approximately 72 species in this genus. The scientific name (which is always written in italics) of each fish species incorporates first the genus and then the species name. This follows the same binomial system used for all animals, as first proposed by Swedish scientist Carl von Linne, commonly known as Linnaeus, in 1753. If the species is not known then the term 'sp.' is used as shorthand for species.

Specalised eyes enable the mudskippers to see both in and out of water.

ELASMOBRANCHII
(Sharks and Rays)

CARCHARHINIFORMES
(Ground Sharks)

Sharks of the order Carcharhiniformes are found in freshwater, brackish and marine environments around the world. There are eight families in the Carcharhiniformes: Carcharhinidae (Requiem Sharks), Hemigaleidae (Weasel Sharks), Leptochariidae (Barbeled Houndsharks), Proscylliidae (Finback Catsharks), Pseudotriakidae (False Catsharks), Scyliorhinidae (Catsharks), Sphyrnidae (Hammerhead, Bonnethead, or Scoophead Sharks) and Triakidae (Houndsharks).

CARCHARHINIDAE
(Requiem Sharks)

Sharks of the family Carcharhinidae are found in marine, brackish and freshwater environments around the world. They have five gill openings, the fifth being located behind the origin of the pectoral fin. Members of this family include small to large sharks with round eyes and internal nictitating eyelids (a transparent inner eyelid that protects and helps keep the eye clean). They have no grooves around the nostrils or barbels, and usually no spiracles (respiratory opening between the eye and the first gill slit in Elasmobranchii, shark spiracles if present are small, but ray spiracles are larger and serve as an opening for water entering the gill chamber). The teeth are usually bladelike, with only one cusp. Reproduction is usually viviparous, with the young born fully developed. This family includes several dangerous species although most prefer to avoid contact with divers.

The name Carcharhinidae is derived from the Greek, 'karcharos' meaning cutting, sharp—and is also the name for one kind of shark—and 'rhinos' meaning nose. There are 12 genera and 50 species in this family.

Carcharhinus sorrah (Muller & Henle, 1839)

Spot-tail Shark

'Yu'

DESCRIPTION: 160 cm. The body is small, and spindle-shaped. The snout is long and rounded, the eyes are large and circular and the teeth are serrated and notched. The second dorsal fin is very close to the caudal fin, with an inter-dorsal ridge present. The first dorsal fin has a thin but conspicuous dark margin, whereas the pelvic fins and upper caudal lobe are plain. The second dorsal fin, pectoral fins and lower caudal lobe all have dark tips. The body is grey or grey-brown above and white below, with a golden-brown sheen on the area between the eyes and the gill slits (in fresh specimens). There is a dark band on the flank extending rearwards to the pelvic fins.

HABITAT: Marine, where it inhabits the continental and insular shelves, primarily near reefs, but can occasionally be found in offshore waters. During the day it is known to linger in the vicinity of the seabed, but remains close to the surface at night.

DIET: Prefers to feed on fish, but will also eat cephalopods and crustaceans.

RANGE: Found throughout the Indo-West Pacific. In Kuching, it is an occasional visitor to Sg. Kuap.

It is viviparous, giving birth to between two and six young per litter. This species is commercially valuable, especially the fins, which are used in the oriental shark fin trade. The liver is used to produce vitamin oil.

RAJIFORMES

(Rays)

Rays of the order Rajiformes are found throughout the marine and freshwater environments of the world. The gill openings are on the ventral side. The anterior edge of the greatly enlarged pectoral fins is attached to the side of the head and anterior to that of the gill openings. The anal fin is absent. The eyes and spiracles (respiratory opening between the eye and the first gill slit, which serves as an opening for water entering the gill chamber) are on the dorsal surface. The body is generally strongly depressed. The jaws are protrusible in most, while the teeth are pavement-like. The water used for breathing is taken in chiefly through the spiracle, rather than the mouth (except for bottom-feeding rays). Almost all rays give birth to live young (some have eggs encased in a horny capsule). The snout may function as an electro-receptive organ (as may be true for all elasmobranches).

The name Rajiformes is derived from the Latin 'raja' meaning ray and 'forma' meaning shape. The ten families in this order are Anacanthobatidae (Smooth Skates), Gymnuridae (Butterfly Rays), Hexatrygonidae (Six-gill Stingrays), Myliobatidae (Eagle and Manta Rays), Plesiobatidae (Deepwater Stingray), Potamotrygonidae (River Stingrays), Rajidae (Skates), Rhinobatidae (Guitar-fishes), Urolophidae (Round Rays) and Dasyatidae.

DASYATIDAE
(Stingrays)

Rays of the family Dasyatidae are found throughout the Atlantic, Indian and Pacific oceans. The stingrays are mostly found in marine environments, but some species dwell in brackish and freshwaters. The side of the head and the anterior margin of the pectoral fin are contiguous. They respire by drawing water through a small hole behind the eye and expelling it through gill slits on the underside of the disc. The dorsal fin is usually absent or, if present, is indistinct. The disc of the body is approximately 1.2 times as broad as it is long. There is no caudal fin. The tail is long and whip-like. Most stingrays have at least one long venomous spine on the tail, which can cause excruciating pain to humans. The largest species grows to about 4 m in length or width. They are viviparous, giving birth to fully developed young.

The name Dasyatidae is derived from the Greek 'dasys' meaning hair, most likely referring to the sting on the tail. There are nine genera and 70 species in this family.

Himantura walga (Müller & Henle, 1841)

Dwarf Whip-ray

'Pari'

Description: 45 cm. The length of the disc is greater than its width, with a long whip-like tail, which is also longer than the body. The snout is produced and pointed. The interorbital space is about three times that of the eye, with the spiracle nearly equal that of the eye. There are between 1 and 3 serrated spines situated one-third of the distance from tail base to the tip, with a series of short spines between the base of the tail and large serrated spines. The middle region of the body is covered with small tubercles. The colour of the dorsal surface of the disc is brownish or dull grey while the ventral surface is whitish.

Habitat: Marine, where it inhabits the benthic substrates of inshore waters.

Diet: Unknown.

Range: Found throughout the Western Pacific. In Kuching, it is found in Sg. Kuap. It has been reported to have originated from India, but due to confusion with *Himantura imbricata*, its westward distribution in the Indian Ocean is unclear.

Dwarf Whip-ray (*Himantura walga*), *'Pari'*.

The spine on the tail of *Himantura walga* gives one of the most painful wounds that can be inflicted by any fish species found in Kuching. Injury usually occurs when people wading in shallow waters accidentally step on the unseen (either buried in the substrate or well camouflaged) ray. When caught in nets, the fishermen usually cut off the tail and spine of any struggling specimens.

ACTINOPTERYGII
(Ray-finned Fish)

Ray-finned fishes, also called bony fishes, differ from the sharks and rays (Elasmobranchii) by the presence of a bone skeleton as apposed to a cartilaginous skeleton, a swim bladder and rayed fins.

OSTEOGLOSSIFORMES
(Bonytongues)

Fishes of the order Osteoglossiformes are mainly freshwater species, although some are known to enter brackish waters. There are seven families in this order: Arapaimidae (Bonytongues), Gymnarchidae, Hiodontidae (Mooneyes), Mormyridae (Elephantfishes), Notopteridae (Featherbacks or Knifefishes), Osteoglossidae (Arowanas) and Pantodontidae (Freshwater Butterflyfish).

NOTOPTERIDAE
(Featherbacks or Knifefishes)

Fishes of the family Notopteridae are found in African and South-East Asian waters and inhabit mostly freshwater, but have also been recorded in brackish water environments. The featherbacks or knifefish have a small quill-like dorsal fin (absent in the genus *Xenomystus*) and a long-based anal fin that merges with the small caudal fin. The combined anal and caudal fins have 100 or more rays and the pelvic fins, when present, are small.

Small specimens are popular with home aquarists and large adults are exhibited in public aquaria. Most species of this family are used for food in their native ranges. Their name comes from the Greek 'noton' meaning back, and 'pteron' meaning wing or fin.

This family contains only three genera and a total of five species. The island of Borneo has one endemic species, *Chitala borneensis,* which is distinctive by its lack of spots along the body.

Chitala ornata (Gray, 1831)
Clown Featherback
'Belida'

Description: 100 cm. It is a distinctive fish deriving its common name 'Knifefish' from its knife-like appearance. The head is small and angular with large eyes and a large blunt mouth that extends past the eyes. From the head, the dorsal region arches strongly, in a hump, before tapering smoothly toward the tail. The ventral region is composed of a long anal fin that merges continuously with the caudal fin.

Habitat: Freshwater, where it inhabits flowing waters of large and medium-sized rivers, but will move into inundated forest during high waters.

Diet: An essentially nocturnal species, it preys on surface-feeding fishes, crustaceans and insects.

Range: Native to Thailand, but due to aquaculture and the aquarium trade has been introduced to most areas in Asia. In Kuching, it is found in the fresher waters of Sg. Sarawak.

Clown Featherback (*Chitala ornata*), *'Belida'*.

Chitala ornata is easily distinguishable from other members of the family by the presence of a row of large dark spots above the base of the anal fin. It differs from *C. chitala* in that it lacks the silver or gold transverse streaks on its dorsum and it differs from *C. blanci* and *C. lopis* in that it lacks a basal pectoral spot.

ELOPIFORMES
(Tarpons and Tenpounders)

The name Elopiformes is derived from the Greek word 'ellops', a kind of serpent, and the Latin word 'forma' meaning shape, combining to translate as a kind of serpent shape. There are only two families in this order: Megalopidae (Tarpons) and Elopidae (Tenpounders).

MEGALOPIDAE
(Tarpons)

Fishes of the family Megalopidae are for the most part marine inhabitants, although they are also commonly found in freshwater, in which they can tolerate low oxygen levels by inhaling air into a lung-like swim bladder. These fish are a popular sport-fish both in fresh and marine waters. The name Megalopidae comes from the Greek word 'mega or megalos' meaning great, and 'pous' meaning foot.

Megalops cyprinoides (Broussonet, 1782)
Indo-Pacific Tarpon or Moonfish
'Bulan'

Description: 55 cm, although commonly caught at lengths of 20 to 30 cm. The body is compressed and covered in large silver scales. The dorsal area is darker in colour than the lighter silver sides and belly. The head is relatively small but dominated by an unusually large eye. The mouth is also large and terminal with the lower jaw projecting beyond the snout. The dorsal fin is situated on the middle of the back and has 16 to 21 rays, of which the last ray is long and filamentous. The anal fin has 23 to 31 rays. The pectoral fins are situated ventrally and the pelvic fins are abdominal with 9 or more rays. The caudal fin is deeply forked.

HABITAT: In the marine environment, *M. cyprinoides* is mainly a pelagic species. Juveniles are commonly found in inner bays, tidal mangrove forests and freshwaters, where they enter rivers, lakes and swampy backwaters. The adults move offshore to breed, possibly throughout the year.

DIET: Feeds on fishes and crustaceans.

RANGE: Found throughout the tropical and subtropical Pacific from the southern coast of Australia north to Hong Kong and India. It is commonly found around the coasts and coastal rivers of Sarawak, and in Kuching is found in Sg. Maong and Sg. Sarawak.

Megalops cyprinoides is diurnally active. It can tolerate low oxygen conditions by 'breathing' air into a lung-like air bladder. It produces recognisable bubbles that can be seen as it rises to the surface to breathe air. This species is commonly cultured in ponds, the fry being sourced from coastal waters. It is a highly popular angling fish.

Indo-Pacific Tarpon or Moonfish (*Megalops cyprinoides*), '*Bulan*'.

CLUPEIFORMES
(Herrings)

Fishes of the order Clupeiformes are mostly plankton feeders, with long and sometimes very numerous gill rakers that serve as straining devices. There are six families in this order: Chirocentridae (Wolf Herrings), Denticipitidae (Denticle Herrings), Engraulidae (Anchovies), Pristigasteridae (Pristigasterids), Sundasalangidae (Sundaland Noodlefish) and Clupeidae (Herrings, Shads, Sardines, Menhadens).

Many of the families and genera belonging to the Clupeiformes are very similar to one another and hence difficult to differentiate. Most species are difficult to keep alive in captivity and obtaining photographs of fresh specimens is extremely difficult as after death the body colour rapidly changes to milky white.

CLUPEIDAE
(Herrings, Shads, Sardines, Menhadens)

Fishes of the family Clupeidae are found throughout tropical and subtropical seas. They are mostly coastal inhabitants, although some species enter freshwater and some species are also anadromous (ascending rivers to spawn). The body is usually elongated and either rounded or strongly compressed. The head is scale-less. The single dorsal fin (when present) is small and near the mid-length of the body while the pelvic fins (when present) are more or less situated below the dorsal fin base. They usually have a keeled abdomen with numerous scutes (hard, thick external plates on the skin derived from scales).

Most Clupeidae are schooling fishes that feed on small planktonic animals. They are important commercial fishes and are processed for food, oil and fishmeal. The name Clupeidae is derived from the Latin 'clupea' meaning sardine, and from the Greek 'clupeus' meaning shield, most likely referring to the scales covering the body of the fish. There are 66 genera and 216 species in this family.

Escualosa thoracata (Valenciennes, 1847)
White Sardine
'Beliak mata'

DESCRIPTION: 10 cm. The body is relatively elongated and compressed. The abdomen is keeled, with approximately 29 scutes. The head is small with moderately large eyes, a small, terminal mouth with a slightly prominent lower jaw. The body is semi-transparent and silvery in colour with a darker silver band running along the mid region from shoulder to caudal fin base. The fins are transparent, but the dorsal and caudal fins have dark margins. The dorsal fin has 13 to 21 soft rays, and the anal fin has 14 to 19 soft rays. It closely resembles some species of the genera *Pellona* and *Illisha*.

HABITAT: Marine and estuarine, commonly forming schools in shallow waters. Juveniles are known to enter the lower parts of rivers, but return later to the sea.

DIET: Plantivorous.

RANGE: Found throughout the Indo-Pacific Ocean. In Kuching, these small sardines are found in the estuarine sections of Sg. Kuap and Sg. Sarawak.

White Sardine (*Escualosa thoracata*), *'Beliak mata'*.

Thryssa hamiltonii (Gray, 1835)
Hamilton's Thryssa
'Bilis'

Description: 27 cm. The body is rectangular and strongly compressed, the keeled abdomen having 23 to 26 scutes. The body is silvery in colour with a darker blue-greenish tinge dorsally, with a noticeable thumbprint-like dark blotch on the shoulder behind the upper part of the gill opening. The caudal fin is forked and green-yellow in colour with a dark margin. The dorsal fin is small, while the anal fin is long with 32 to 39 soft rays. The tip of the snout is just above the level of the eye centre.

Habitat: Marine, a pelagic schooling fish which inhabits inshore waters, regularly entering estuaries.

Diet: Feeds on prawns, polychaetes, copepods and amphipods.

Range: Found throughout the Indo-Pacific. In Kuching, it is found in the estuarine sections of Sg. Kuap.

Hamilton's Thryssa (*Thryssa hamiltonii*), *'Bilis'*

Anodontostoma chacunda (Hamilton, 1822)
Chacunda Gizzard Shad
'Kuasi'

Description: 17 cm, although usually caught at lengths between 10 and 14 cm. The body is oval in shape, deep and compressed. The dorsal region is a slightly metallic, bluish-brown colour and silvery below. There is a distinctive thumbprint-like blotch behind the operculum.

HABITAT: Marine and estuarine, where it inhabits mostly coastal waters but ascends rivers up to the tidal zone.

DIET: Feeds on diatoms, radiolarians, molluscs, copepods, and crustaceans.

RANGE: Found throughout the Indo-Pacific. In Kuching, it is commonly found in Sg. Kuap and also in the lower section of Sg. Sarawak.

Chacunda Gizzard Shad (*Anodontostoma chacunda*), *'Kuasi'*.

PRISTIGASTERIDAE
(Illisha, Herring and Tardoore)

Fishes of the family Pristigasteridae are found throughout tropical waters of the Atlantic, Pacific, and Indian oceans and in the freshwaters of South America and South-East Asia. They are externally distinguishable from other sardines (Clupeidae) by their long anal fins, which have at least 30 fin rays. The pelvic fins are very small or absent. The head is naked and the body is elongated and compressed, with scutes along the belly. The dorsal fin is short and situated mid-body, the pectoral fin is moderate to large.

The species classified under the genera *Illisha*, *Pellona* and *Pristigaster* were formerly under the family Clupeidae. There are nine genera and 37 species in this family.

Illisha elongate (Bennett, 1830)
Slender Shad or Elongate Illisha
'Beliak Mata'

Description: 40 cm, although usually caught at lengths between 12 and 16 cm. The body is compressed and oval in shape, with a strongly keeled abdomen having between 34 and 42 scutes. The eye is large, and the lower jaw is projecting. The dorsal fin origin is about midpoint on the body or a little behind, and the anal fin (43 to 53 soft rays) origin is a little behind the dorsal fin base. The caudal fin is deeply forked. The lateral line is absent. The dorsal region is bluish or bluish-green in colour and silvery below. The dorsal, anal and caudal fins have a dark margin.

Habitat: A warm-water coastal inhabitant, entering estuaries where it is known to ascend rivers up to the tidal zone.

Diet: Unknown.

Range: Found throughout the Indo-Pacific. In Kuching, it is commonly found in Sg. Kuap and also in the lower section of Sg. Sarawak.

During cooler years, this species will not reach the northern part of its range. It is used in Chinese medicine.

ENGRAULIDAE
(Anchovies)

Fishes of the family Engraulidae, derived from the Greek word for anchovy, are coastal inhabitants found throughout the Atlantic, Indian and Pacific oceans. They are usually elongated and strongly compressed with an oblong keel, notable for having scutes.

They are mostly schooling fishes, feeding chiefly on zooplankton but some species are known to be piscivorous. A new name 'Engraulididae' (replacing Engraulidae) has been proposed for this family. The Engraulidae comprises 16 genera and 139 species.

Coilia dussumieri (Valenciennes, 1848)
Goldspotted Grenadier Anchovy
'Gonjeng'

Description: 20 cm. The body is elongated and compressed, tapering toward the slender, pointed tail. The body colour can be either a dark orange bronze or silver, which pales towards the ventral regions. This species is characterised by rows of golden or pearly spots on the ventral region of the body (most notable in juveniles and smaller specimens). These are light organs located beneath the scales. They also occur along the edge of the lower jaw and on the cheek and gill covers. The eyes are large and rounded and the snout is blunt. The belly is rounded before the pelvic fins, and has 12 to 15 keeled scutes situated from just behind the pectoral fin base to the anus. The pectoral fin has approximately 6 long filamentous rays, which reach the start of the anal fin, and 9 to 11 (rarely, 8) branched fin rays. The anal fin is long (80 soft rays) and converges with the caudal fin.

It can be differentiated from *C. macrognathus* by the presence of the golden spots along the body.

Habitat: Coastal and estuarine waters.

Goldspotted Grenadier Anchovy (*Coilia dussumieri*), *'Gonjeng'*.

Diet: Feeds on copepods, prawns and fish larvae.

Range: Found throughout the tropical waters of the Indo-Pacific Ocean. In Kuching, it is a regular visitor to Sg. Kuap.

Coilia dussumieri possibly enters estuaries to breed. It is a commercially valuable species.

Setipinna breviceps (Cantor, 1849)
Short-head Hairfin Anchovy
'Empirang'

Description: 24 cm. The body is elongated and compressed with a small head and a large mouth with a lower jaw which extends beyond the tip of the snout. The body is a bright copper colour with a black margin on the caudal and pectoral fins. The dorsal fin has a semi-translucent white margin. The upper ray of the pectoral fin is elongate, forming a long black filament which extends past the dorsal fin. The anal fin is long with 59 to 64 soft rays. There are 27 to 32 scutes between the isthmus and the anus. The mouth is strongly oblique. Its very low gill raker count and the characteristic shape of its head and jaws distinguish it from all other species of *Setipinna*.

Habitat: Marine and estuarine, and will ascend into fresh water.

Diet: Unknown.

Short-head Hairfin Anchovy (*Setipinna breviceps*), *'Empirang'*.

Range: Found in the coastal waters of the Indian Ocean and western central Pacific. In Malaysia, it is only found in the waters around Sarawak, and In Kuching, it is found in Sg. Kuap.

Setipinna breviceps is a commercially valuable species.

Setipinna taty (Valenciennes, 1848)
Scaly Hairfin Anchovy

Description: 16 cm. The body is elongated and strongly compressed with a small head and relatively large jaw. The colour is silvery with a slight dark olive-green tint to the dorsal region. The caudal fin is forked and is pale translucent yellow in colour with a dark margin. The anal fin has 45 to 55 soft rays. There are 32 to 40 keeled scutes from the isthmus to the anus along the ventral region. The upper ray of the pectoral fin is elongated and forms a long filament, which extends past the dorsal fin.

Habitat: Coastal and estuarine waters, and known to enter rivers.

Diet: Unknown.

Range: Found throughout the coastal waters of the Indo-West Pacific. In Malaysia, it has only been recorded from Sarawak's coastal and estuarine waters. In Kuching, *S. taty* is found in the estuarine sections of Sg. Kuap.

Setipinna taty is usually found in schools.

Scaly Hairfin Anchovy (*Setipinna taty*).

CYPRINIFORMES
(Carps)

Fishes of the order Cypriniformes are one of the most numerous of the orders, including all the goldfish and goldfish-like families. This order of fishes is predominantly found in freshwater habitats and, to a lesser extent, in brackish water environments.

The name Cypriniforme is derived from the Greek 'kyprinos' meaning goldfish. There are six families in the order Cypriniformes: Balitoridae (River Loaches), Catostomidae (Suckers), Cobitidae (Loaches), Cyprinidae (Carps and Minnows), Gyrinocheilidae (Algae Eaters) and Psilorhynchidae.

CYPRINIDAE
(Carps and Minnows)

Fishes of the family Cyprinidae are found from North America (northern Canada to southern Mexico), Africa, and throughout Eurasia. They are freshwater inhabitants although some are found in brackish water environments. This is the most dominant of the freshwater families with over 200 genera and over 2000 species. Many of the genera and most species found in Borneo are endemic to South-East Asia.

All cyprinids have toothless oral jaws but well-toothed pharyngeal jaws (pharynx) with 1 to 3 rows, each with a maximum of eight teeth. Despite a lack of jaw teeth, they have a diverse array of feeding habits. There are a few larger species, the maximum length being 2.5 m, but the majority of species reach only 5 cm or less. Cyprinidae are generally non-guarders, but in some species the males build nests and/or protect the eggs.

Leptobarbus hoevenii (Bleeker, 1851)
Mad Barb
'Sultan'

Description: 100 cm, although usually caught at around 20 to 30 cm in length. The body is elongated, slender and dorsally compressed. The scales are large and have a darker colouration above, while silvery below. The ventral fins are a bright red and the caudal fin has a red margin, which is more vivid on the bottom fork. The dorsal fin has 9 soft rays, while the anal fin has 8 soft rays. Juveniles and smaller specimens have a black longitudinal stripe which fades with growth.

The native species to Borneo is *Leptobarbus hosii*. The number of scales between the dorsal fin and the lateral line can differentiate these two species; *L. hosii* has five and a half scales while *L. hoevenii* has four and a half scales.

Habitat: A freshwater species introduced to Borneo, it has successfully adapted to the local environment, often out-competing the local species. It prefers freely flowing rivers and streams and will move into flooded areas.

Diet: Eats parenchyma and seeds of the chaulmoogra tree (*Hydnocarpus* species) that fall into the streams, after which it has been reported to become intoxicated and to behave strangely.

Range: Native to Indo-China but has been introduced to Borneo through aquaculture. In Kuching, it is found in Sg. Maong and the fresher waters of Sg. Sarawak.

The flesh of this species is highly esteemed throughout much of Asia and this species is very important to the aquaculture industry.

Mad Barb (*Leptobarbus hoevenii*), *'Sultan'*.

Puntius orphoides (Valenciennes, 1842)

Javaen Barb

'Bantak buin'

Description: 25 cm. The body is dorsally compressed and silvery in colour. The head is small with a pointed, terminal mouth and 4 barbels. This species has conspicuous markings comprising a black spot on the caudal peduncle, a small, vertically elongated black spot below the dorsal origin, a narrow dark band from the upper part of the gill opening to the pectoral base and a red blotch on the operculum. The fins are mostly transparent, with the ventral fins reddish in colour. The forked caudal fin has a red outer margin and black edges, the colouration being most vivid on the lower lobe. There are often longitudinal rows of faint black spots along scale rows. The dorsal fin has 4 spines and 8 soft rays. The anal fin has 2 spines and 5 soft rays.

Habitat: A freshwater species, it inhabits many river environments, from large to small rivers, canals and flooded plains.

Diet: Unknown.

Range: Found throughout Thailand, Peninsular Malaysia, Indonesia and Borneo. In Kuching, it is found in Sg. Maong and the fresher waters of Sg. Sarawak.

Puntius orphoides is commonly captured from the wild for the ornamental fish trade.

Javaen Barb (*Puntius orphoides*), *'Bantak buin'*.

Barbonymus gonionotus (Bleeker, 1850)
Java Barb
'Lampan jawa'

Description: 41 cm. The body is strongly compressed, the back is elevated, and the dorsal profile arched. The head is small and pointed, with a small terminal mouth. The barbels are very minute or rudimentary and, in the case of the upper barbels, may be entirely absent. The colour when fresh is silvery white, sometimes with a golden tint. The dorsal and caudal fins are grey to grey-yellow; the anal and pelvic fins light orange with reddish tips. The pectoral fins are pale to a light yellow. The dorsal fin has 4 spines and 8 soft rays and the anal fin has 3 spines and 6 to 7 soft rays.

Habitat: A freshwater species that prefers standing water habitats instead of flowing waters, and inhabits the flooded forest during high water periods.

Diet: Feeds on plant matter (e.g. leaves, weeds, *Ipomea aquatica* and *Hydrilla* sp.) and invertebrates.

Range: Once confined to the Indo-China regions, it is now found throughout Asia. This expansion is due to the species being a favourite of the aquaculture trade, hence its introduction to Borneo. In Kuching, it is found in Sg. Maong and in the fresher waters of Sg. Sarawak.

Java Barb (*Barbonymus gonionotus*), *'Lampan jawa'*.

Barbonymus gonionotus was once known as *Puntius gonionotus*, and as escapees from culture installations it has become established in rivers and forms the basis for capture fisheries on several South-East Asian islands. It is known to be useful for cropping excessive vegetation in reservoirs.

Puntius sealei (Herre, 1933)
Three-spot Barb

DESCRIPTION: 17 cm. It is a silver coloured fish with a slight reddish bronze tinge and has 4 black round or oval spots situated along the middle of the body, above the lateral line. There is also a small dark spot at the anterior base of the dorsal fin, and another at the base of the anal fin. All the fins are a clear translucent colour, while the lower pectoral rays, the entire pelvic, the first two thirds of the anal fin and the lower caudal lobe have an orange tint. The last dorsal ray is bony and serrated behind. The pelvic ray is longer than the adjacent branched ray, thus forming a filament in specimens smaller than 6 cm long. In young specimens, the end of the first anal ray has a black spot.

HABITAT: Freshwater, where it inhabits small, clear or slightly turbid waters.

Three-spot Barb (*Puntius sealei*).

Diet: Feeds on plants, insects and crustaceans.

Range: Native to Borneo. In Kuching, it is found in Sg. Maong and the fresher waters of Sg. Sarawak.

Rasbora tornieri (Ahl, 1922)
Yellowtail Rasbora
'Seluang'

Description: 17 cm. The body is elongated, slender and compressed with a small conical shaped head. The eyes are large and the mouth is small, pointed and terminal. It has a wide darkish stripe, which has a metallic sheen, down the length of the body. It is generally easily recognised by the broad black margin on the bright yellow caudal fin although in many of the specimens observed from the rivers In Kuching, the tail is rather dull.

Habitat: A freshwater species that inhabits streams, canals and ditches in lowland floodplains, where it tends to dwell at the surface.

Diet: Feeds on exogenous insects.

Range: Found throughout Indo China, Malaysia and Indonesia. In Kuching, it is found in Sg. Sarawak.

Rasbora tornieri spawns in rivers and ponds.

Yellowtail Rasbora (*Rasbora tornieri*), *'Seluang'*.

Ctenopharyngodon idella (Valenciennes, 1844)

Grass Carp

'Ikan China'

Description: 1.5 m. The body is a dark bronze above and a white silvery bronze below, while the fins are all a darker brown than the body. The dorsal fin has a total of 3 spines and 7 to 8 soft rays. The anal fin has 3 spines and 7 to 11 soft rays. There are no barbels. The snout is very short, its length less than or equal to the eye diameter.

Habitat: A freshwater species that inhabits lakes, ponds, pools and backwaters of large rivers. It prefers large, slow-flowing or standing water bodies with vegetation.

Diet: Feeds on higher aquatic plants and submerged grasses but is also an opportunistic scavenger and will eat detritus as well as insects and other invertebrates.

Range: Found throughout Asia and Eastern Europe. In Kuching, it is found in Sg. Sarawak.

Ctenopharyngodon idella has been widely transported around the world via the parasitic nematode *Bothriocephalus opsarichthydis*. Several countries have reported adverse ecological impacts after introduction and have considered it as a pest because of the damage done to submerged vegetation.

Grass Carp (*Ctenopharyngodon idella*), *'Ikan China'*.

It is one of the world's most important aquaculture species and also one used for weed control in rivers, fishponds and reservoirs. The species spawns on riverbeds where the current is very strong.

SILURIFORMES
(Catfish)

Fishes of the order Siluriformes are found in all types of habitat from marine to freshwater. The most common distinguishing features associated with this order are: (1) barbels, which are usually found in four pairs and are important in detecting food, (2) serrated spines usually in the dorsal fin and frequently in the pectoral fins, many of which have toxins and (3) the body, which is usually naked or covered in bony plates.

The name Siluriformes comes from the Greek word 'silouros', or 'silurus' meaning sly. This probably refers to its sly, slinking motion and shape, both of which are similar to sharks. This is a fairly large order with 33 families.

BAGRIDAE
(Bagrid or Freshwater Catfish)

Fishes of the family Bagridae are found throughout Africa and Asia where they inhabit both fresh and brackish water environments. Their bodies are naked and elongate and the head usually has four pairs of well-developed barbels. The dorsal and pectoral fins have a well-developed serrated spine. Many species of this family are valued both as food fish and aquarium fish. It is a large family with 30 genera and 210 species.

Mystus gulio (Hamilton, 1822)
Long-whiskered Catfish
'Nyingan baong'

Description: 46 cm. The body is elongate and dark coloured with a flattened white underbelly. The head is short, blunt and slightly depressed, with 4 pairs of well-developed barbels, the longest of which reaches backwards to the tip of the caudal fin. The fins are a dusky colour and semi-translucent. The caudal fin is forked and often has a slightly darker margin.

Habitat: Mostly a brackish water inhabitant but may enter and live in freshwater, preferring larger rivers with mud or clay substrates.

Diet: Unknown.

Range: Generally found around countries bordering the Indian Ocean, but may have been introduced to other Asian countries as a game fish. In Kuching, it is mostly found in Sg. Sarawak and occasionally in Sg. Maong.

Mystus gulio often congregates in schools of 10 to 25 individuals.

A close-up of the head of *Mystus gulio,* showing the well-developed upper and lower barbels common to the Bagrid catfish.

Long-whiskered Catfish (*Mystus gulio*), *'Nyingan baong'*.

Hemibagrus nemurus (Valenciennes, 1840)
Asian Red-tail Catfish
'Baong baong'

DESCRIPTION: 65 cm. The body is elongate with a flattened silvery white underbelly. The head is short, blunt and slightly depressed with a rugose skull roof. The eyes are large and slightly raised above the surface of the head. The body colour ranges from a light grey to brown, often with a greenish sheen. The fins are grey with a violet tint. The dorsal fin has 2 spines (the second of which is prominent) and 7 soft rays. The pectoral fin spines are serrated along the inner edge and smooth in front. The anal fin has 10 to 13 soft rays. The caudal fin is forked and the tip of the upper lobe extends, forming a short black filament in fresh specimens. There are 4 pairs of well-developed barbels; the nasal barbels extend to or beyond the eyes, the maxillary barbels reach the anal fin, the mandibulary barbels reach beyond the base of the pectoral fins and the mental barbels reach two-thirds to three-quarters the distance between their base and the insertion of pectoral fins.

Asian Red-tail Catfish (*Hemibagrus nemurus*), *'Baong baong'*.

Habitat: A freshwater species that can extend downstream to the tidal brackish waters of larger rivers.

Diet: Feeds on exogenous insects, aquatic insect larvae, shrimps, other crustaceans and fishes.

Range: Found throughout Indo-China and Sumatra, Java and Borneo. In Kuching, it is mostly found in Sg. Sarawak and occasionally in Sg. Maong.

ARIIDAE
(Sea Catfish)

Fishes of the family Ariidae are some of the hardest to identify with over 120 species. Member of this family are mostly marine, but many enter freshwater and some are even restricted to freshwater. They are found in tropical and subtropical waters.

The Sea Catfish are elongate, with a naked body, a broad, rounded and depressed head and strong bony plates known as 'head shields' located on the top of the head extending to the base of the dorsal fin. They are characterised by one pair of maxillary (extending from the lateral part of the upper jaw) barbels and two pairs of mandibulary (extending from the lower jaw) barbels. There are no nasal barbels. One important feature used to identify Ariidae is the bands and patches of teeth found on the palantine. Ariidae have strong, posteriorly serrated leading spines in both the pectoral and dorsal fins, which are covered in poisonous mucus that can inflict extremely painful wounds. They all have forked caudal fins.

Ariidae generally prefer a muddy substrate where they use their barbels to find food. They are oral incubators, with the male carrying the relatively large eggs in the mouth until hatching.

Ten species have been found in the rivers around Kuching, with only five species confidently identified. The rest have been described and tentatively given names. The family Ariidae is currently under review.

Arius sagor (Hamilton, 1822)

Sagor Catfish

'Badukang'

DESCRIPTION: 45 cm. One of the more distinctive species of Ariidae: the bony head shields are strongly granulated and the pre-dorsal plates have a distinctive butterfly shape. The maxillary barbels extend to the middle region of the pectoral fin. The body is a dark grey-bronze colour, with faint metallic stripes present on the dorsal region and flank, paling to a white

Sagor Catfish (*Arius sagor*), *'Badukang'*.

belly. The fins are all a dusky black colour with the pelvic fin having a narrow white margin.

The upper palantine tooth patches have a narrow band behind the lips and four oval patches—two outer and two inner, the outer ones being slightly larger than the two inner ones. These patches almost join together, forming a band that follows the same curve as the narrow band behind the lips.

Habitat: Coastal waters where it favours river mouths and estuaries, ascending into freshwater of the upper tidal zone.

Diet: Feeds on invertebrates and small fishes.

Range: Found throughout the Indo-West Pacific and South-East Asia. In Kuching, it is found in Sg. Kuap and occasionally in the brackish waters of Sg. Sarawak.

Arius truncatus (Valenciennes, 1840)

Marine Catfish

'Lundu la-la'

Description: 42 cm. The body is elongate with a narrow and depressed head. The head has bony plates with a marked central ridge and the eyes are small. The body is a bright metallic gold colour, slightly darker above, with a yellow belly. The fins are all

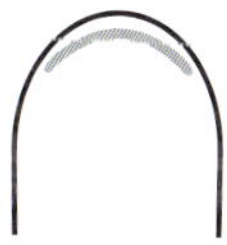

Marine Catfish (*Arius truncatus*), *'Lundu la-la'*.

a similar transparent golden colour with a slightly dusky margin on the dorsal and adipose fins and a slightly paler margin on the ventral and caudal fins.

The palantine teeth have a single curved band on both the upper and lower jaws, just behind the lips. Some specimens have two, symmetrical, very small indistinct patches on the roof of the mouth.

Habitat: A coastal inhabitant that regularly enters estuaries and the lower courses of rivers.

Diet: Feeds on fish and crustaceans.

Range: Found throughout the Western Central Pacific. In Kuching, it is found mostly in Sg. Sarawak in large schools and less commonly in Sg. Kuap.

Arius truncatus usually schools in large groups which are regularly caught in gillnets.

Arius maculatus (Thunberg, 1792)

Spotted Catfish

'Lundu'

Two distinct forms were found in Kuching of what appear to be *Arius maculatus*. Both have been described and compared.

Description: 80 cm. The usual length for both varieties caught in Kuching is between 15 and 40 cm. Both the catfish species described below have the same basic body shape and colouring consisting of a metallic grey body and pale belly. The head shield is rugose with a deep and long median fontanelle groove. The dorsal and pectoral spines have relatively strong venom, with 7 soft rays in the dorsal and 16 to 30 soft rays in the anal fin. The dorsal fin has a dark margin and the adipose fin has a black spot.

The most important feature apart from these is the palantine roof patches, which are nearly similar if not identical. There is a narrow band behind the lips and on the interior part of the roof of the mouth there are two symmetrical teardrop-shaped patches pointing back towards the throat. These are covered in small knobble-like protrusions.

The two pictures below show different variations that have been found in the rivers around Kuching. *Arius maculatus* sp. 1 has a long filament, which extends from the first dorsal spine for a length that can range from a few millimetres to touching the adipose fin. This variation in length may be associated with the current health of the fish. During sampling, the majority of those caught with filaments were from more estuarine areas while those without filaments were mostly caught further upriver.

(Top). *Arius maculatus* sp. 1 (long filament).
(Bottom). *Arius maculatus* sp. 2 (no filament).

Arius maculatus sp. 2 is identical in appearance to *Arius maculatus* sp. 1 described above, but with no sign of a filament on the first dorsal spine.

Habitat: Coastal inshore waters and estuaries.

Diet: Feeds on invertebrates and small fishes.

Range: Normally, *Arius maculatus* is found throughout the Indo-West Pacific and the Indo-Australian Archipelago (excluding Australia). In Kuching, *Arius maculatus* sp. 1 is generally found in Sg. Kuap, while *Arius maculatus* sp. 2 is found usually in large schools in Sg. Sarawak.

Both specimens have been caught in large schools.

Arius microcephalus (Bleeker, 1855)
Marine Catfish
'Lundu'

Description: 60 cm, but specimens usually caught between 10 and 35 cm. It is very similar in appearance to both of the *Arius maculatus* species described above: the same basic body shape and colouring comprising a metallic grey body and pale belly. The dorsal fin has a dark margin and the adipose fin has a black spot.

Marine Catfish (*Arius microcephalus*), *'Lundu'*.

The palantine roof patches are also extremely similar, possibly knobblier, if not identical. There is a narrow band behind the lips and on the interior part of the roof of the mouth there are two symmetrical teardrop-shaped patches pointing back towards the throat. These are covered in small knobble-like protrusions.

This species is similar to *Arius maculatus* sp. 2, but with no filament. The distinction from *Arius maculatus* is the mouth which, by comparison, is much narrower and with thicker, pursed-like lips.

HABITAT: Inhabits the tidal reaches of rivers and the inshore and brackish waters of coastal areas.

DIET: Thought to eat molluscs and crustaceans.

RANGE: Found in east Thailand and Borneo. In Kuching, it is found mostly mixed in with schools of *A. maculatus* sp. 2 in Sg. Sarawak and is less regularly found in Sg. Kuap.

A close-up of the mouth of *Arius microcephalus* with its pursed-like lips. A section of palantine teeth can also be seen behind the top lip.

Arius nenga (Hamilton, 1822)
Marine Catfish
'Lundu'

Description: 30 cm. It is very similar in appearance to *Arius truncatus*, but less elongate and the head is not so depressed. The body is a metallic bronze colour with a slightly paler belly. The dorsal fine has a darker margin and a long black filament. The adipose fin has a distinct black spot. All other fins are a paler version of the body.

The palantine tooth patches have a narrow band behind the lips and two large patches directly behind the first band form two rather angular teardrop shapes. These are situated diagonally, with one side running parallel to the first band, the gap between them forming two sides of a triangle.

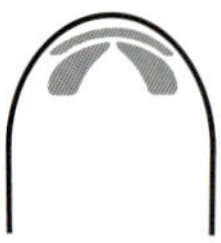

Habitat: Marine, prefers turbid inshore waters such as estuaries, known to enter upstream to the tidal reaches of rivers.

Diet: Unknown.

Range: Found throughout the Indo-Pacific Ocean. In Kuching, it is found in large schools in Sg. Kuap.

Marine Catfish (*Arius nenga*), *'Lundu'*.

Arius sp. 1 (Red-eye)
Marine Catfish
'Lundu'

DESCRIPTION: Specimens were found around 10 to 20 cm in length. This species is very similar to *A. nenga*. The body is also a metallic bronze colour with a slightly paler belly. The dorsal fin has a darker margin but no long black filament and the adipose fin is dusky coloured. All other fins are a paler version of the body. The eyes are smaller and rather distinctively reddish in colour, which is different from most of the other Ariidae that have been found in rivers around Kuching.

The palantine tooth patches have a narrow band behind the lips. Two large patches directly behind the first band, rather than forming two angular teardrops, form more of a diamond shape, one side of which runs parallel to the first band. The two patches are situated further apart than in *A. nenga*, and the triangle between the patches is more oblong.

HABITAT: Presumably coastal and estuarine.

DIET: Unknown.

RANGE: Global range unknown. In Kuching, it is found in mixed schools with *A. nenga* in Sg. Kuap.

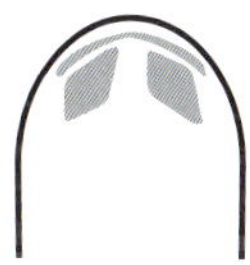

Marine Catfish (*Arius* sp. 1 (Red-eye)), *'Lundu'*

Arius sp. 2 (Broad-headed)
Marine Catfish
'Lundu'

DESCRIPTION: Specimens were found to be between 15 and 30 cm in length. This species is similar in appearance to *A. maculatus*. The body is a metallic grey colour on the upper side and silvery white below. The dorsal fin has a dusky margin and the adipose fin has a black spot. The head is shorter than other Ariidae species with a strongly blunted mouth.

The palantine tooth patches are a narrow band directly behind the lips with a pair of very long narrow oval patches directly behind and parallel to the first band.

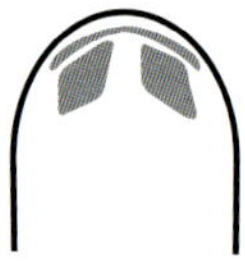

HABITAT: Presumably coastal and estuarine.

DIET: Unknown.

RANGE: Global range unknown. In Kuching, it is found in the estuarine sections of Sg. Kuap.

Ketengus typus (Bleeker, 1847)
Bigmouth Sea Catfish
'Lundu'

DESCRIPTION: At 24 cm, *Ketengus typus* is one of the smaller species of Ariidae. The body is a bright gold colour with all fins the same colour or paling to a translucent white. The head is small and strongly blunted while

Bigmouth Sea Catfish (*Ketengus typus*), *'Lundu'*.

the mouth is wide and extends behind the eye. The lips are large and very fleshy. The bony part of the roof of the mouth is small with no palantine teeth apart from the first band directly behind the lips. All three pairs of barbels are very short, equal to or less than eye diameter.

Habitat: A coastal and brackish water inhabitant, entering upstream to the tidal reaches of rivers.

Diet: Feeds on invertebrates and small fishes.

Range: Found throughout the coasts of Thailand, West Malaysia, the north-western islands of Indonesia, southern Borneo, this is possibly the first record for Sarawak. In Kuching, it is found occasionally in Sg. Kuap.

Osteogeneiosus militaris (Linnaeus, 1758)

Military Catfish

'Lundu'

Description: 35 cm. *O. militaris* is more easily distinguishable from other Ariidae species by (a) the absence of the two pairs of mandibulary barbels and (b) the maxillary barbels being very stiff. The top of the body is a metallic grey colour, while the lower body and undersides are a yellow bronze colour. The adipose fin has a dark spot. The head has an almost perfectly straight line from the snout to the base of the dorsal fin. The eyes are small and directly above the corner of the mouth.

Military Catfish (*Osteogeneiosus militaris*), *'Lundu'*.

The palantine teeth are fairly distinctive, with a single band directly behind the lips and two patches that resemble inverted speech marks.

Habitat: A coastal species that inhabits estuaries and river mouths.

Diet: Feeds mainly on invertebrates and small fishes.

Range: Found throughout the Indo-West Pacific and South-East Asia. In Kuching, it is found in Sg. Sarawak and Sg. Kuap.

PLOTOSIDAE
(Eel-tailed Catfish)

Fishes of the family Plotosidae are marine although many species enter estuaries and even freshwater. They are found throughout the Indo-West Pacific (Japan to Australia and Fiji). The body is easily recognised by the long eel-like shape with a pointed or bluntly rounded tail. The head is depressed and bluntly rounded with no bony plates. They usually have four pairs of barbels: nasal, maxillary and two pairs of mandibulary barbels. The dorsal and pectoral spines contain a potent venom.

The Plotosidae are benthic feeders, using their barbels to detect food. The juveniles usually congregate in tight schools, which only break up when they become adult. The Plotosidae contains nine genera and 32 species.

Plotosus canius (Hamilton, 1822)
Eel-tailed Catfish
'Sembilang'

Description: 1.5 m, although usually caught between 50 to 80 cm. The body is a plain dusky-brown colour, which may display a banded pattern at night. The dorsal fin has a black tip. It is distinguished from the similar species *Plotosus lineatus* by its long nasal barbels, which can reach past the eyes. The maxillary barbels reach the pectoral fins, while the mandibulary barbels are short. The spines associated with anterior fins have a potent venom, which can be extremely painful.

Habitat: Coastal waters, especially estuaries and lagoons, but it sometimes can be found up river in near fresh waters.

Diet: Feeds mainly on crustaceans, molluscs and fish.

Range: Found throughout the Indo-West Pacific. In Kuching, it is found in Sg. Kuap and occasionally in the brackish waters of Sg. Sarawak.

Juveniles of *Plotosus canius* are often caught in small groups with the cast net. Fishermen are careful to cut the spines before handling these fish.

SILURIDAE
(Sheatfish)

The Eurasian Siluridae is one of the most distinctive and well-defined families of the catfish. The body is naked and compressed and the head is depressed. There are a pair of maxillary barbels, which are usually elongate, one or two pairs of mandibulary barbels, and no nasal barbels. The dorsal fin, when present, is spineless, the pelvic fins are small or absent and the anal fin is very long (41 to 110 rays). There is no adipose fin.

The Sheatfishes are a primarily freshwater family with over 100 species and 21 species known from Borneo. This family can reach a massive length of 5 m (*Silurus glanis*). The name Siluridae is derived from the Latin word 'silurus' meaning a kind of fish.

Kryptopterus cryptopterus (Bleeker, 1851)
Sheatfish
'Lajong'

Description: 20 cm. The body is elongated and slightly bowed with a single dorsal fin that is very small and indistinct. The caudal fin is forked and the transparent anal fin is long, running almost the entire length of the body, while the pectoral fins are small and transparent. The head is small and triangular with a terminal mouth, and the eyes are relatively large, set low behind the corners of the mouth. The pair of long maxillary barbels reach half the body length. The organ cavity is slightly enlarged and translucent. The body is a light golden brown on the dorsal region above the head, while the rest of the body is a pale almost translucent pearl colour with a pinkish sheen.

Habitat: Generally inhabits large turbid rivers, streams and canals.

Diet: Feeds on fish, prawns, insects and their larvae.

Sheatfish (*Kryptopterus cryptopterus*), *'Lajong'*.

Range: Found throughout Asia. In Kuching, it is found in the fresher waters of Sg. Sarawak.

Spawning in *Kryptopterus cryptopterus* occurs during the early part of the rainy season, and the young later move toward the seasonally flooded habitats.

LORICARIIDAE
(Armoured Catfish)

Fishes of the family Loricariidae live primarily in freshwater. They are originally from South America, but as a favourite of the aquaria industry they have been inadvertently introduced to many tropical countries.

The body is covered in bony plates. The mouth is inferior, with or without conspicuous barbels. There are papillae on the ventral lip, which have evolved into a sucker-like mouth that allows the fish to cling to surfaces. The adipose fin is absent or, when present, with a spine at the anterior border. The name Loricariidae is derived from the Latin 'lorica' meaning corselet. This family has approximately 550 species.

Liposarcus pardalis (Castelnau, 1855)
Armoured Catfish
'Ikan DBKU'

DESCRIPTION: 40 cm. The body is elongated and somewhat flattened, especially the belly. The head is flattened with a sub-terminal mouth and two pairs of short barbels. From the snout the profile slopes up and arches towards the dorsal region. The eyes are medium and widely spaced with a pair of nasal organs with tufts. The body is mottled with a dark and light brown maze-like pattern that extends onto all the fins. The body is covered

A close-up of the head of *Liposarcus pardalis,* clearly showing its sucker-like mouth attached to the glass bottom of an aquarium.

Armoured Catfish (*Liposarcus pardalis*), *'Ikan DBKU'*.

in hard bony mail-like armour, with small spines that form three similar, parallel rows down its length. The dorsal and pectoral fins have a strong predominate spine. The dorsal fin is large and extends from the middle of the pectoral fin to the caudal peduncle. The pelvic and pectoral fins are well developed and large. The anal fin is small, situated to the posterior side of the mid-body. The caudal fin has two long predominate rays on either side and is slightly concave, the ventral ray being longer than the upper.

HABITAT: Inhabits the bottom of fast-flowing streams as well as slow and highly polluted waters.

DIET: Omnivorous, but feeds mainly on algae and benthic detritus.

RANGE: Native to South America, from the lower, middle and upper Amazon River basin, but has been released via the aquarium trade and has become firmly established in many Asian counties. In Kuching, it is found in Sg. Maong and the fresher waters of Sg. Sarawak.

Liposarcus pardalis is a very hardy solitary species that can to some extent breathe air and thus is easily transported between different locations and is able to survive in slow, sluggish waters where oxygen levels are poor. An almost identical species, *Liposarcus disjunctivus*, has a marbled belly instead of the spotted belly found in *L. pardalis*. It is highly likely that both species have been released and become well established in the rivers of Kuching.

PANGASIIDAE
(Shark Catfish)

Fishes of the family Pangasiidae are found in southern Asia, from Pakistan to Borneo, inhabiting freshwater and, occasionally, brackish water environments. Their bodies are elongated and smooth-skinned with a flattened, depressed head. There are usually only two pairs of short barbels. The dorsal and pectoral fins are well developed and have strong, serrated spines. The adipose fin is small, while the anal fin is fairly long. The name Pangasiidae is derived from the Vietnamese name for, presumably, a similar type of fish. There are two genera and 21 species in this family.

Pangasius hypophthalmus (Sauvage, 1878)
Sutchi Catfish
'Patin'

Description: 1.3 m. The body is elongated and slightly compressed, with a flattened, depressed head. The body colour is a uniform metallic grey, darker above and fading to a narrow silver ventral region. The fins are darker

Sutchi Catfish (*Pangasius hypophthalmus*), *'Patin'*.

with even darker grey or black margins. The caudal fin is widely forked with rounded lobes. Juveniles have two black stripes running the length of the body, one along the lateral line and one below. The dorsal fin has 6 branched rays.

Habitat: A freshwater species that inhabits large rivers, migrating upstream to breed.

Diet: Omnivorous, feeding on fish and crustaceans as well as vegetable debris.

Range: An Asian species well documented from the Mekong Basin. It is an important agricultural species and has been introduced to Borneo from West Malaysia and Thailand. In Kuching, it is found in Sg. Maong and the fresher waters of Sg. Sarawak.

Pangasius hypophthalmus was previously known as *Pangasius sutchi.*

CLARIIDAE
(Airbreathing Catfish)

This distinctive family of airbreathing catfish is found in the freshwater habitats of both Africa and Asia. The Clariidae are easily recognisable by their extremely elongated bodies, the dorsal (spineless) and anal fins often extending the entire length of the body and a rounded caudal fin giving them an oval appearance. The head is blunt and depressed with small eyes and four pairs of well-developed barbels. Airbreathing is accomplished with a labyrinthic organ arising from the gill arches, giving its nickname 'labyrinth catfishes'. They also have the nickname 'walking catfishes' due to the ability of some species to travel short distances over land. The name Clariidae comes from the Latin word 'clarias' meaning shining. There are 13 genera and 100 species in this family.

Clarias nieuhofii (Valenciennes, 1840)
Slender Walking Catfish
'Keli'

Description: 50 cm. The body is elongated and dark grey in colour, fading to a pale cream belly before the start of the anal fin. There is a distinguishing series of white spots set in vertical rows that extend down the entire length of the body, and also an indistinct line of spots running along the mid-flank of the body. The head is blunt and depressed with very small eyes and four sets of barbels, three of which reach back past the short pectoral fin. The dorsal and adipose fins are fused to form one long fin that stretches from just behind the end of the pectoral fin all the way to the caudal, with 87 to 106 soft rays. The anal fin covers just over half the body length with 69 to 95 soft rays. The pectoral fin is small but still has a strong serrated spine.

HABITAT: A freshwater species that inhabits lakes and streams.

DIET: Unknown.

RANGE: Found throughout South-East Asia. In Kuching, it prefers the stagnant water of Sg. Maong.

Slender Walking Catfish (*Clarias nieuhofii*), *'Keli'*.

Clarias gariepinus (Burchell, 1822)

North African Catfish

'Keli'

DESCRIPTION: 1.7 m. The body is generally a dark, greyish-black colour above and a creamy-white below, with a fairly distinct black longitudinal band on either side of the ventral surface of the head. Juveniles (less than 9 cm long) have a similar colouring, except that the black bands on the head are absent; these bands become more distinct as the fish grows. The upper surface of the head is coarsely granulate in adult fishes and smooth in the young. The eyes are small. The dorsal fin has 61 to 80 soft rays and the anal fin has 45 to 65 soft rays. The pectoral fin has a spine, which is serrated only on its outer side. All barbels are a similar length as the head with the maxillary barbels, usually somewhat longer.

Habitat: A freshwater species that prefers to inhabit quiet waters such as lakes and pools, but may also occur in fast-flowing rivers and rapids.

Diet: Mostly a nocturnal benthic feeder foraging on a wide variety of prey, including insects, plankton, invertebrates, fish, rotting flesh, plants, and has also been known to catch young birds.

Range: Originally found throughout Africa and northern Asia. In Kuching, it is found mostly in Sg. Maong and occasionally the fresher waters of Sg. Sarawak.

Clarias gariepinus is a highly tolerant and hardy species that can adapt to extreme environmental conditions. This has enabled it to adapt well after introduction to the rest of Asia and southern Europe, where it has caused some adverse ecological impacts. The presence of an accessory-breathing organ enables it to breathe air either when very active or under very dry conditions. During drought conditions it can remain in the muddy substrates of ponds, occasionally gulping air through its mouth. If necessary, it can leave the water at night using its strong pectoral fins and spines in search of land-based food or to move into the breeding areas through very shallow pathways.

North African Catfish (*Clarias gariepinus*), *'Keli'*.

ANGUILLIFORMES

(Eels and Morays)

Fishes of the order Anguilliformes can be found in marine, freshwater and brackish-water environments. The name Anguilliformes is derived from the Latin word 'anguila' meaning eel or eel-shaped. There are 15 families in this order, varying from the freshwater eels to the well-known Conger, Moray and Snake Eels.

MURAENESOCIDAE

(Pike Congers)

Fishes of the family Muraenesocidae are found in marine, brackish and freshwater environments throughout the tropical Atlantic, Indian and Pacific oceans. The name Muraenesocidae comes from the Latin word 'muraena' meaning moray eel, and the Greek word 'esox' meaning nursery of salmon. There are four genera and eight species in this family.

Muraenesox bagio (Hamilton, 1822)

Dagger-tooth Pike Conger

'Andong'

Description: 200 cm. It has a long robust and eel-shaped body and a small and narrow head. The eyes are large and the snout is long with a set of relatively large and sharp teeth. The upper body is a brown copperish colour with a slight metallic sheen, which fades in the ventral region to a white belly. The dorsal fin (which has 47 to 49 rays before reaching the anus) extends from the top of the head to converge with the small caudal fin. The anal fin is much shorter, but also joins with the caudal fin. Both the dorsal fin and the anal fin are darker with a black margin. The lateral line extends the entire length of the body with 33 to 39 lateral line pores before reaching the anus.

Habitat: Marine, where it inhabits coastal and estuarine waters.

Diet: Feeds on benthic fishes and crustaceans.

RANGE: Found in the Indo-West Pacific. In Kuching, it is mostly caught in Sg. Kuap.

Muraenesox bagio is mostly a nocturnal species. When captured alive it is difficult to handle due to the large teeth on the end of a long, writhing body.

Dagger-tooth Pike Conger (*Muraenesox bagio*);
a close-up of the head shows a formidable set of teeth.

SYNBRANCHIFORMES
(Spiny Eels)

The name Synbranchiformes comes from the Greek word 'syn' meaning together, or adhesion, and 'brangchia' meaning gills, which together translate as fused gill shape. Families include Chaudhuriidae, Mastacembelidae (Spiny Eels) and Synbranchidae (Swamp Eels).

MASTACEMBELIDAE
(Spiny Eels)

Fishes of the family Mastacembelidae are mostly freshwater inhabitants found in tropical and subtropical areas from Africa, through Syria to Malaysia and China. The name Mastacembelidae is derived from the Greek word 'mastax', meaning bite or mouthful, and 'emballo' meaning to throw oneself into. Together they translate as 'a fish that likes to throw itself into the bite'. The common name 'Spiny eel' is derived from the row of razor-sharp spines along the dorsal fin. Almost all species of this family have a fleshy appendage on the tip of the snout, i.e. an extended elephant-like nose. They live in various freshwater habitats. This family contains four genera and 67 species.

Mastacembelus erythrotaenia (Bleeker, 1850)
Spiny Eel, Fire Eel
'Tilan'

Description: 100 cm. It is one of the most striking of the spiny eels, with usually one or more bright red lateral lines along the dark brown body and spots along the tail, while the anal, pectoral and dorsal fins frequently have a red margin. These markings vary among individuals depending on health and age. The markings tend to be a yellow colour in juveniles, changing to red in larger adults. The body is very elongate with the dorsal and anal fins converging with the caudal. The head is small with a fleshy trunk-like appendage on the tip of the snout. As the common name suggests, the dorsal fin has numerous small razor-sharp spines.

Spiny Eel or Fire Eel (*Mastacembelus erythrotaenia*); the close-up of the head shows the fleshy trunk-like appendage at the tip of the snout.

Habitat: A freshwater species that generally inhabits slow-moving rivers and floodplains, hiding amongst plants and objects or buried in the river mud.

Diet: Feeds on benthic insect larvae, worms and some plant material, and has also been seen to ambush small prawns.

Range: Found throughout South-East Asia. In Kuching, it is found in the fresher waters of Sg. Sarawak.

Mastacembelus erythrotaenia is a favourite in the aquarium trade, but is becoming scarce in its natural habitat.

BELONIFORMES
(Needlefish)

Beloniformes, as the Greek word 'belone' meaning needle suggests, are fish that have moderately to highly elongated body shapes. Beloniformes contains five families: Adrianichthyidae (Ricefish), Belonidae (Needlefish), Exocoetidae (Flyingfish), Hemiramphidae (Halfbeaks) and Scomberesocidae (Sauries).

BELONIDAE
(Needlefish)

Belonidae are predominantly marine inhabitants, but some species also inhabit brackish and freshwater. They are found worldwide, both in tropical and temperate waters. Belonidae are long and slender shaped fish with both upper and lower jaws extended into long beaks armed with sharp needle-like teeth. They are generally green or blue on the back and silvery-white on the lower sides and belly. These fish are surface dwellers and the colouring helps protect them against predators from above and below. Freshwater species tend to be smaller, reaching only six or seven centimetres in total length, while some marine species may attain two metres.

Belonidae is a carnivorous family, feeding mainly on small fish, which they catch sideways in their beaks. Needlefishes are well known for leaping along the water surface; this behaviour may be to escape predators, but it is well documented that they are attracted to lights at night, often leaping into fishing boats. Eggs have entangling tendrils to enable them to stick together in clumps and to cling to objects in the water. Belonidae contains 10 genera and 34 species.

Strongylura strongylura (van Hasselt, 1823)
Spottail Needlefish, Garfish
'Todak'

Description: 40 cm. The body is elongate, slender and sub-cylindrical. The body colour is silvery with an olive green tinge along the dorsal profile and silvery-white on the belly. The head is narrow with extremely long, slender 'beak-like' jaws, of which the lower jaw is slightly longer than the upper jaw, and the eyes are large. There is a shallow, narrow median grove along the head between the eyes. It is easy to distinguish from other Belonidae species by the presence of a dark spot at the base of the caudal fin. The dorsal and anal fins are opposite, on the posterior end of the body. The dorsal fin is smaller than the anal fin, with 12 to 15 rays and 15 to 18 rays respectively. The pelvic fins are abdominal and slightly posterior to the mid-point of the body. The caudal fin is fan-shaped to rounded. The lateral line runs down from pectoral fin origin and parallel with the ventral margin of the body. The body has small scales, which are cycloid and easily detached.

Habitat: Marine, where it inhabits coastal areas, mangrove-lined lagoons and may occasionally enter freshwater.

Diet: Feeds mainly on small fishes, especially of the clupeid order, and crustaceans.

Spottail Needlefish or Garfish (*Strongylura strongylura*); a close-up of the head shows the beak is filled with small but numerous sharp teeth.

Range: Found throughout the Indo-West Pacific. In Kuching, it is found in the estuarine sections of Sg. Kuap.

In *Strongylura strongylura* the prolonged jaws are not present in juveniles, but develop as the fish grows.

HEMIRAMPHIDAE
(Halfbeaks)

Hemiramphidae are known from the Atlantic, Indian and Pacific oceans, where most species live in marine and brackish waters although there are a few freshwater species. The name Hemiramphidae comes from the Greek word 'hemi' meaning half and 'rhamphos' meaning beak or bill, hence the common name Halfbeak. As this name suggests, the lower jaw is a long needle-like beak, while the upper jaw is shorter (less than half the length of the lower) and triangular.

Hemiramphidae attach their eggs to algae in shallow waters. They inhabit the surface waters where they have an omnivorous diet including algae, floating seagrasses, zooplankton, crustaceans and small fishes. The Hemiramphidae contains 14 genera and 117 species.

Zenarchopterus sp.
Halfbeak, Feathered River-Garfish
'Nyulung'

Description: Members of the *Zenarchopterus* genus have an elongate and sub-cylindrical body, generally darker on the dorsal profile and pale silver to white on the belly. The head is small, with an extremely long lower

Juvenile of the Halfbeak, *Zenarchopterus* sp. 2.

jaw and a shorter, triangular upper jaw. The eye is relatively large. The dorsal and anal fins are found on the posterior end of the body close to the caudal fin. The anal fin is usually larger than the dorsal fin. The pelvic fin is situated to the posterior side of the mid-body. There are several species of *Zenarchopterus* in Malaysian waters and identification is difficult unless one studies the modification of the anal fin rays in mature males. *Zenarchopterus* are sexually dimorphic (males and females are either different in structure or in colour) and reproduction is carried out with internal fertilisation.

Habitat: The *Zenarchopterus* are generally found in the shallow waters around mangroves, estuaries and the tidal zone of rivers.

Diet: Possibly feeds on small terrestrial and aquatic insects, insect larvae and other small invertebrates.

Range: In Kuching, members of this genus are found at the river edges of both Sg. Kuap and Sg. Sarawak.

Halfbeak (*Zenarchopterus* sp. 1), 'Nyulung'. The close-up of the head clearly shows the much smaller upper beak that gives rise to the common name 'half-beak'.

Current taxonomy states there are five genera and about 50 sexually dimorphic species in the estuarine and fresh waters of the Indo-West Pacific region in the sub-family Zenarchopterinae (Meisner & Collette, 1998). However, morphological and molecular evidence suggests that the subfamily Zenarchopterinae needs to be raised to family level for the five genera and about 54 species of internally-fertilising Halfbeaks, as they are more closely related to Belonidae and Scomberesocidae than to the "normal" Halfbeaks of the rest of the family Hemiramphidae.

SYNGNATHIFORMES
(Pipefish and Seahorses)

Fishes of the order Syngnathiformes are found in marine to freshwater environments. The bodies are elongate, and encased in a series of bony rings and most species have small mouths with tube-shaped snouts. The name Syngnathiformes comes from the Greek word 'syn' meaning with or together and 'gnathos' meaning jaw; this refers to the tube-like mouth.

The Syngnathiformes includes five families: Aulostomidae (Trumpet-fish), Centriscidae (Snipefish and Shrimpfish), Fistulariidae (Cornetfish), Solenostomidae (Ghost Pipefish) and Syngnathidae (Pipefish and Seahorses).

SYNGNATHIDAE
(Pipefish and Seahorses)

Fishes of the family Syngnathidae are found throughout the Atlantic, Indian, and Pacific oceans (mostly in warm temperate to tropical regions). They are mainly marine inhabitants, though some species are found in brackish and freshwater environments.

The Syngnathidae have an elongate body enclosed by a series of bony rings and generally have a single dorsal fin (though in some species this disappears in the adults). The anal fin is small, and there are no pelvic fins. The caudal fin is small and in some species is lacking altogether, while the caudal peduncle is prehensile in the Seahorses. The gill openings are very small.

The Syngnathidae are usually limited to shallow waters where they feed on minute invertebrates sucked into their tubular snouts. The males have a brood pouch into which the females lay eggs and where the eggs are fertilised and incubated.

Doryichthys boaja (Bleeker, 1851)
Long-snouted Pipefish
'Kerek' or *'Undok-undok'*

DESCRIPTION: At 41 cm, *Doryichthys boaja* is one of the largest freshwater species of pipefish. It has a conspicuous colour pattern with alternating blue and brown bars; there are 22 to 24 rings on the trunk and 32 to 38 rings on the tail. The dorsal fin is found on the posterior region (towards the tail) of the body and has 43 to 65 soft rays.

HABITAT: It inhabits rivers and streams.

DIET: Feeds on small crustaceans, worms and insects.

RANGE: Found throughout Asia. In Kuching, it is found in the less brackish waters of Sg. Kuap.

Long-snouted Pipefish (*Doryichthys boaja*), *'Kerek'* or *'Undok-undok'*.

As with all Syngnathidae, the males of *Doryichthys boaja* carry the eggs in a brood pouch under the tail until they hatch.

PERCIFORMES
(Perch-like fish)

Perciformes is the largest order of the fishes with 156 families. Most families in many suborders are not currently definable in terms of shared derived characteristics and thus may not be monophyletic (descended from a common ancestor). The majority of Perciformes are marine shore fishes, but there are still a large number (approximately 2000 species, e.g. cichlids) that normally occur only in freshwater and another 2200 species that occur in freshwater for at least part of their life history.

The name Perciformes is derived from the Greek word 'perke' meaning perch. This is the group that contains the perch-shaped fishes.

CHANNIDAE
(Snakeheads)

The Channidae is a freshwater family, found in tropical Africa and southern Asia. Their bodies are elongate, with long dorsal and anal fin bases. They have a specialised organ which enables them to breathe air. Some species have become important in aquaculture and rice-fish farming, and hence have been widely introduced in both Asia and southern Europe. It is thought that most snakeheads produce pelagic (surface) eggs. A nest built and guarded by one or both parents protects these eggs.

The name Channidae comes from the Greek 'channe' meaning anchovy. *Channa* sp. used to be *Ophicephalus* sp. There are 13 genera and 49 species in this family.

Channa striata (Bloch, 1793)
Snakehead Murrel
'Haruan'

Description: 100 cm. The body is elongate and sub-cylindrical, the scales are large, the head is depressed and the caudal fin rounded. The dorsal surface and sides are dark and mottled with a combination of black and ochre while the belly is white. The head is large, reminiscent of a snake's head with a deeply gaping, fully-toothed mouth. The dorsal fin has 38 to 43 soft rays and the anal fin has 23 to 27 soft rays.

Habitat: A freshwater species that inhabits ponds, streams and rivers, preferring stagnant and muddy waters.

Diet: Feeds on fish, frogs, snakes, insects, earthworms, tadpoles and crustaceans.

Range: It has spread throughout most of Asia, including Indo-China, Malaysia, Philippines and Indonesia. In Kuching, it is found in Sg. Maong and occasionally the fresher waters of Sg. Sarawak.

Channa striata can survive on stored fat through dry spells or seasons by burrowing into the mud bottom of lakes, canals or swamps, but only if the skin and air-breathing apparatus remain moist.

This is possibly the main marketed fish in Thailand, Indo-China and Malaysia, the firm white flesh being almost bone-free and the heavy dark skin (usually sold separately) good for soup.

Snakehead Murrel (*Channa striata*), *'Haruan'*.

Channa micropeltes (Cuvier, 1831)

Giant Snakehead

'Toman'

DESCRIPTION: At 130 cm, *Channa micropeltes* is the largest of the family Channidae. The body is elongate and sub-cylindrical with very large scales; the head is depressed and the caudal fin rounded. The dorsal surface

The juvenile colouration of the Giant Snakehead (*Channa micropeltes*) with the two longitudinal stripes (adults have a single stripe). The close-up shows the charachteristic 'snake-head'.

and sides are dark and mottled with a combination of black and ochre while the belly is white. The head is large, reminiscent of a snake's head, and the mouth is deeply gaping, fully toothed. The body of the adult has a broad, dark longitudinal stripe along the side from eye to caudal while that of juveniles (from roughly 30 mm to 45 cm in length) has two black longitudinal stripes with the area in between a bright orange or scarlet colour.

Habitat: A freshwater species usually associated with deep-water inhabiting large streams and canals, which have standing or slowly flowing water.

Diet: Predatory, feeding mainly on fish but also some crustaceans.

Range: Found throughout most of Asia, including Indo-China, Indonesia and Borneo. In Kuching, it is found in Sg. Maong and the fresher waters of Sg. Sarawak.

Channa micropeltes is known to be an aggressive guarder of its nest and eggs, and has been known to attack humans who approach too close. It is also known to be destructive to fish populations, preying on various species and sizes and killing more than it can eat.

OSPHRONEMIDAE
(Gouramies)

Fishes of the family Osphronemidae are freshwater species found throughout Indo-China to Malaysia and Indonesia (the Malay Archipelago). The common feature of the majority of Gouramies is the elongated ray or filament that extends from the pelvic fin. Some species are oral brooders and others build bubble nests.

Many of the *Betta* sp., *Parasphromenus* sp., *Sphaerichthys* sp. and *Trichopterus* sp. were named as members of Belontiidae, but were later grouped together under Osphronemidae. This family has also been called Polyacanthidae. There are four subfamilies: Belontiinae (Combtail Gouramies), Macropodinae (Siamese Fighting Fish or Paradisefish), Osphroneminae (Giant Gouramies) and Trichogastrinae (Gouramies). The name Osphronemidae comes from the Greek 'osphra' meaning smell and 'nema' meaning filament, suggesting these fish use their long filaments to smell their environment. There are 13 genera and 29 species in this family.

Osphronemus goramy (Lacepède, 1801)
Giant Gourami
'Kaloi'

Description: 70 cm and can weigh up to 5 kg. The body is deep and compressed. The head is small and triangular with large eyes and a sharply pointed terminal mouth. The dorsal and, especially, the anal fins are large, the bases extending to the caudal peduncle. The distal margin of the anal fin reaches a point equal to the distal margin of caudal fin. The caudal fin is large and rounded. The dorsal fin has 12 to 14 spines and 10 to 13 soft rays. The anal fin has 9 to 13 spines and 18 to 21 soft rays. The first ray of the pelvic fin is extended into a thread-like filament reaching backwards to or beyond the hind margin of the caudal fin. Juveniles have 8 to 10 dark vertical bars,

Giant Gourami (*Osphronemus goramy*), *'Kaloi'*.

which fade and disappear in adults to become a drab brown colour. As they grow larger, some are known to develop a horn on the forehead; these are seen as the most prestigious and lucky.

Habitat: A freshwater species commonly found in medium to large rivers and stagnant water bodies including sluggish flowing canals and the drainage ditches of padi fields. It favours swamps, lakes and rivers, and will enter flooded forests.

Diet: Omnivorous, feeding on both plants and animals, such as some aquatic weeds, fish, frogs, earthworms, and is sometimes known to scavenge from fresh carcasses.

Range: Probably native to Sumatra, Borneo, Java, the Malay Peninsula, Thailand and Indo-China, but has been introduced to several other countries via aquaculture. In Kuching, it is found in Sg. Maong and the fresher waters of Sg. Sarawak.

Osphronemus goramy can breathe directly from moist air and is thus able to survive in still and low oxygenated waters. It also means it can be kept alive for long periods out of water, making it easy to transport fresh, either to ponds or market. It is well known in the aquarium industry as well as for consumption.

Trichogaster trichopterus (Pallas, 1770)
Three-spot Gourami
'Sepat'

Description: At 15 cm, *Trichogaster trichopterus* is one of the smaller Gouramies found in the Kuching rivers. The body is oval shaped and ventrally compressed. The head is small and the mouth terminal and pointed. The colour is a pale brown with irregular dark marks on the shoulders and numerous narrow irregular oblique bars on the body. The opercles and thorax are golden yellow. The long anal fin often has a bright orange-red margin with numerous small orange spots along its length. The caudal fin is shallowly forked or concave with rounded lobes and has numerous small orange-brownish spots. The dorsal fin is small, with 6 to 8 spines and 7 to 10 soft rays, most of which are branched. The anal fin has 9 to 12 spines and 30 to 38 soft rays, most of which are branched. The scales are of moderate size, irregularly arranged, and the lateral line is curved. There are two characteristic black spots, one in the middle of the body and one at the base

A wild specimen of the Three-spot Gourami (*Trichogaster trichopterus*); the three spots are indistinct compared to captive specimens bred for the aquarium trade.

of the caudal fin. The photo depicts the wild species of the Three-spot Gourami, in which the three spots are not so obvious as the inbred variety.

Habitat: A freshwater species, it favours marshes, swamps and shallow, sluggish or standing-water bodies that have an abundance of aquatic vegetation.

Diet: Feed on zooplankton, crustaceans and insect larvae.

Range: Found throughout Indo-China, Indonesia and Borneo, and has been introduced to much of the rest of Asia, including the Philippines and Bali. In Kuching, it is found in Sg. Maong and the fresher waters of Sg. Sarawak.

As with most of the Gouramies, *Trichogaster trichopterus* can breathe directly from the air, enabling it to survive in low oxygenated conditions such as stagnant water, ponds, canals and sluggish rivers. It is marketed fresh, but also commonly seen in the aquarium trade.

Trichogaster microlepis (Günther, 1861)
Moonlight Gourami
'Sepat'

Description: 13 cm. The body is slightly elongate and ventrally compressed. The head is small and the mouth is terminal and pointed; the lower jaw is longer than the upper. The dorsal profile is rounded, but with a slight concave depression from the snout to the top of the head. The body is a plain silvery-blue with no markings. The dorsal fin, situated mid-body, has 34 soft spines. The anal fin has a long base extending from the pelvic fin area to the caudal peduncle. The caudal fin is shallowly forked with rounded lobes. The pelvic filament extends backwards to the tip of the caudal fin.

Habitat: Freshwater.

Diet: Feeds on zooplankton, crustaceans and aquatic insects.

Range: Originally native to the Mekong, in Cambodia and Viet Nam, and the Chao Phraya basin, but has been introduced to much of Asia and has even been introduced to Colombia via escapees from aquarium rearing ponds. In Kuching, it is found in Sg. Maong and the fresher waters of Sg. Sarawak.

Moonlight Gourami (*Trichogaster microlepis*), *'Sepat'*.

As with most of the Gouramies, *Trichogaster microlepis* can breathe directly from the air and is thus able to survive in low oxygenated conditions such as stagnant water, ponds, canals and sluggish rivers with an abundance of aquatic vegetation. It is a very popular aquarium species, its main method of introduction. It is also marketed fresh.

Trichogaster pectoralis (Regan, 1910)
Snakeskin Gourami
'Sepat'

DESCRIPTION: 25 cm. The body is slightly elongate and ventrally compressed. The head is small and the mouth is terminal and pointed. The dorsal profile is rounded, but with a slight concave depression from the snout to the top of the head. The body has numerous dark, slightly diagonal cross-bands, which are not always distinct, and the presence of an irregular black strip along the side reaching from the eye to the middle of the caudal fin base. The dorsal fin is relatively central, with 7 to 8 short spines and 10 to 11

Snakeskin Gourami (*Trichogaster pectoralis*), *'Sepat'*.

longer soft rays. The anal fin starts mid-body, below the dorsal fin, with 9 to 12 spines and 33 to 38 soft rays, but becomes larger and more rounded towards the caudal fin. The caudal fin is very shallowly forked with rounded lobes. The pelvic fin filament extends backwards to the tip of the caudal fin.

Habitat: Freshwater.

Diet: Generally feeds on aquatic plants.

Range: Native to the Mekong basin in Laos, Thailand, Cambodia and Viet Nam and also the Chao Phraya basin, but has been introduced to much of the rest of Asia. In Kuching, it is found in Sg. Maong and the fresher waters of Sg. Sarawak.

Trichogaster pectoralis can breathe directly from the air thus enabling it to survive in low oxygenated conditions such as stagnant water, ponds, canals and sluggish rivers with an abundance of aquatic vegetation. It is a favourite in the aquaculture industry, as well as for market consumption.

ANABANTIDAE
(Climbing Gouramies)

Fishes of the family Anabantidae are found from Africa and India to the Philippines. This is a freshwater family although occasionally found in brackish waters. The name Anabantidae comes from the Greek 'anabas' or 'aoristo' or 'anabainein' meaning to climb. There are three genera and 30 species in this family.

Anabas testudineus (Bloch, 1792)
Climbing Perch
'Puyu'

Description: 25 cm. The body form can vary, from oval and compressed to elongate and sub-cylindrical, with age and the amount of food consumed. It can be dark greenish, grey or pink in body colour, paling towards the ventral regions. The head has longitudinal stripes ventrally, the posterior margin of opercle has a small dark spot and the iris of the eye is golden reddish in colour. The body is patterned with rows of vertical bands. The scales are large and arranged regularly. The dorsal fin starts above the

Climbing Perch (*Anabas testudineus*), *'Puyu'*.

pelvic fin with 16 to 20 small sharp spines (sharp but non-poisonous) and extends to a short-based rounded lobe just above the caudal peduncle and has 7 to 10 soft rays. The anal fin is the mirror image of the soft part of the dorsal fin with 9 to 11 spines and 8 to 11 soft rays. The caudal fin in rounded.

HABITAT: Generally inhabits canals, lakes, ponds, swamps and estuaries, usually in water bodies with dense vegetation.

DIET: Feeds on macrophytic vegetation, shrimps and fish fry.

RANGE: Found from Indo-China to the Wallace Line in Sulawesi, but may be more commonly distributed around Asia. In Kuching, it is found in Sg. Maong.

Anabas testudineus is famous for its ability to 'walk' between bodies of water. It is one of the hardiest species of fish and has a most advanced accessory air-breathing organ, enabling it to survive several days or even weeks without food and out of water provided the surrounding air is moist. It can therefore tolerate extremely unfavourable conditions such as drought, when it will remain buried in mud until water returns, and the low levels of oxygen associated with turbid, stagnant waters.

CICHLIDAE
(Cichlids)

Fishes of the family Cichlids were originally found in Central and South America, the West Indies, and from Africa to the coasts of southern India. But due to both aquaculture and the aquarium trade many species have been widely introduced throughout the tropics and sub-tropics. The body shape is variable, but generally it is moderately deep and compressed. The Cichlids are well documented for their breeding and guarding behaviour, with three generalised forms of parental care: mouth brooding, substrate brooding, and substrate brooding of eggs followed by mouth brooding of the fry. Cichlidae is one of the most species-rich families in freshwaters worldwide, with estimates of species numbers ranging from 1300 to 1900.

Because of their varied behavioural characteristics and attractive colours, cichlids are a favourite of the aquarium trade. Practically all genera and more than half of the species have been kept in aquaria at some time, while the larger species are used as food fish, many being important in aquaculture.

Oreochromis niloticus niloticus (Linnaeus, 1758)
Nile Tilapia
'Tilapia'

Description: 60 cm. The body is oval and compressed with a large triangular shaped head. The eyes are large and the mouth is terminal and pointed. The colour is a pinkish grey with indistinct vertical bands along the body. The dorsal fin base is long, extending from the top of the head to the caudal fin, with 16 to 18 spines and 12 to 13 soft rays. The anal fin is shorter and has 3 spines and 9 to 11 soft rays. The pelvic fins are large and pointed, situated below and slightly posterior to the transparent pectoral fins. The most distinguishing characteristic of this species is the presence of 7 to 12 regular vertical stripes throughout the depth of the caudal fin. There are many hybrids of *Oreochromis niloticus niloticus* and many coloured variations. Among these coloured variations is the pink coloured Tilapia, which is probably mostly crossed between *Oreochromis aureus* and *O. niloticus*.

Nile Tilapia (*Oreochromis niloticus niloticus*), *'Tilapia'*.

Habitat: A freshwater species which can inhabit a wide variety of habitats such as rivers, lakes, sewage canals and irrigation channels.

Diet: Feeds on phytoplankton or benthic algae.

Range: Introduced to Asia for its simplicity in breeding and growing in aquaculture conditions. Its distribution is now almost worldwide as it is reared in aquaculture stock as the flesh is tasty, and it is a hardy species that is very easy to breed. In Kuching, it is found in Sg. Maong and the fresher waters of Sg. Sarawak.

The varied-coloured specimens are favourites both in aquaria and with restaurants. It is popularly believed that these varied-coloured specimens have a superior taste although it is unlikely that there is any difference in the flavour. *Oreochromis niloticus niloticus* is mainly diurnal, and is a potential pest as it will out-compete and out-breed most native species.

Etroplus suratensis (Bloch, 1790)
Green Chromide

Description: 40 cm. The body is oval in shape and compressed, the dorsal profile from the mouth to the back is strongly arched with a small bump opposite the eye. The head is slightly triangular with large eyes and a medium-sized terminal mouth. The colouring is striking, with numerous gold flecks covering the entire body apart from the face and the ventral region of the head. There are indistinct alternating pink and green vertical bands along the body. The dorsal fin has the longer soft ray portion near the caudal fin and is a mirror image of the anal fin. The caudal fin has a flat terminal margin and is a clear translucent colour.

Habitat: In the wild, it inhabits both brackish and freshwaters including large rivers, reservoirs, lagoons and estuaries.

Diet: Feeds on filamentous algae, plant material and insects.

Range: Thought to be originally from India and Sri Lanka, where it is popular aquaculture species, but may have been introduced or spread to many countries throughout Asia, including Borneo. In Kuching, it is found in Sg. Sarawak and Sg. Kuap.

Green Chromide (*Etroplus suratensis*).

DREPANEIDAE
(Sicklefishes)

Fishes of the family Drepaneidae are found from the Indo-West Pacific and West Africa. The body is deep and laterally compressed. The name Drepaneidae is derived from the Greek, 'drepane' meaning sickle. There is only one genus containing three species. There are two species common to the Indo-Pacific and Malaysia: *Drepane punctata* and *D. longimana*, which can be differentiated by the presence or absence of spots on the upper half of the body.

Spadefish or Spotted Sicklefish (*Drepane punctata*), *'Baharu-baharu'*.

Drepane punctata (Linnaeus, 1758)

Spadefish, Spotted Sicklefish

'Baharu-baharu'

DESCRIPTION: 50 cm, although usually caught at around 25 cm. It has a deep and strongly compressed angular body, with an approximate diamond shape. The lateral line is strongly curved towards the dorsal region. The colour is generally silvery with greenish tinge above, and a series of 4 to 11 (usually 8) vertical spots on the dorsal region. The pectoral fin is long and pointed reaching the caudal peduncle.

HABITAT: Inhabits various inshore habitats such as sand or muddy bottoms, reefs, estuaries and harbours.

DIET: Feeds on benthic invertebrates and fishes.

RANGE: Found throughout the Indo-Pacific. In Kuching, it is found in Sg. Kuap.

Drepane punctata is very similar to *D. longimana*; the only difference being that the series of spots are faint solid bands in *D. longimana*. These two species are mainly distinguishable only by their colour and may possibly be the same species.

TRICHIURIDAE
(Cutlassfish or Ribbonfish)

Fishes of the family Trichiuridae are found in brackish and marine environments from the Atlantic, Indian, and Pacific oceans. As the name 'Ribbonfish' suggests, the body is extremely elongated and strongly compressed. The jaws usually have fanglike teeth. The dorsal fin is very long, extending over the entire length of the body; the soft-rayed part is usually longer than the spinous part, with the two parts separated by a notch in some species.

The name Trichiuridae comes from the Greek 'thrix' meaning hair and 'oura' meaning tail, translating as Hairtail; this refers to the long tapering hair-like tail which in some species has a small fork on the tip.

There are nine genera in this family containing some 32 species. There are two very similar species in the waters around the coasts of Sarawak: *Lepturacanthus savala* and *Trichiurus lepturus*. The photograph opposite has been identified as *L. savala*, but the colouring appears a little different from the normal descriptions.

Lepturacanthus savala (Cuvier, 1829)
Savalani Hairtail or Ribbonfish
'Timah'

Description: 100 cm. The fish tapers to a long ribbon-like tail. The colour is a steely blue with metallic reflections, darker along the dorsal fin base to a silvery white belly. There are 3 to 4 dorsal spines and 110 to 120 soft dorsal rays. The pelvic and caudal fins are absent and the anal fin is reduced to about 75 spinules. The pectoral fin has a slight black margin. The lateral line runs from the top of the head and down along the ventral profile. Most specimens are pale to white on the posterior region, and the tip of the jaws is black.

Habitat: Coastal waters.

Diet: Feeds on a variety of small fishes and crustaceans, chiefly on prawns and species of *Setipinna*, *Anchoviella* and *Harpodon*.

Range: Found throughout the Indo-West Pacific, from India and Sri Lanka to South-East Asia, north to China, and south to New Guinea and northern Australia. In Kuching, it is found in Sg. Kuap.

Lepturacanthus savala appears as a smaller version of one of the long dragon-like creatures of myths and legend, with some formidable teeth. This species is often found near the water surface at night.

Savalani Hairtail or Ribbonfish (*Lepturacanthus savala*), *'Timah'*.

SILLAGINIDAE
(Smelt-whitings)

Fishes of the family Sillaginidae are brackish and marine inhabitants, found throughout the Indo-West Pacific. They are elongated cigar-like fish with small, pointed mouths. They have two dorsal fins and a long anal fin. The Silliganids are a coastal family, feeding mainly on benthic or epibenthic organisms. Some species are important in estuarine aquaculture.

The name Sillaginidae comes from the Greek 'syllego' meaning to meet. There are three genera and 31 species in this family.

Sillago sihama (Forsskål, 1775)
Silver Sillago
'Ucus'* or *'Bulus-bulus'*, or *'Ubi'

Description: 30 cm. The head is conical with a small pointed mouth. The entire body is silvery with a slight brownish tinge dorsally. The dorsal fin has 12 to 13 spines and 20 to 23 soft rays. The anal fin has 2 spines and 21 to 23 soft rays.

Habitat: Beaches, sandbars, mangrove creeks and estuaries. It will enter the lower reaches of tidal rivers.

Silver Sillago (*Sillago sihama*), *'Ucus'* or *'Bulus-bulus'* or *'Ubi'*.

Diet: Adults are omnivorous, feeding mainly on polychaete worms, small prawns and shrimps, amphipods and even seaweed and corals. The larvae and juveniles are found in pelagic waters feeding on plankton.

Range: Found throughout the Indo-West Pacific. In Kuching, it is found in Sg. Kuap and the more brackish waters of Sg. Sarawak.

Sillago sihama is commonly found in schools and, when disturbed, will rapidly bury itself in the sand, diving headfirst until it completely disappears.

MULLIDAE
(Goatfishes)

Fishes of the family Mullidae are usually associated with sandy and muddy substrates of tropical coastal waters of the Atlantic, Indian, and Pacific oceans, though occasionally found in brackish waters. Goatfish are recognisable by the presence of two long barbels beneath the chin. These barbels contain chemosensory organs and are used to probe the sand or mud for benthic invertebrates or small fish. The males waggle their barbels vigorously during courtship.

The name Mullidae comes from the Latin 'mullus' meaning soft. There are six genera and 55 species in this family.

Upeneus sulphureus (Cuvier, 1829)
Yellow or Sulphur Goatfish
'Nangka'

Description: 28 cm. The body is moderately elongate. The first dorsal fin has 8 spines and the second dorsal fin is made up of 8 soft rays. The dorsal fins have 2 to 3 olive bars with black or dark brown tips; these are more pronounced in the first dorsal fin. The anal fin has 1 spine and 7 soft rays, and, similar to the pelvic and pectoral fins, is pale yellow in colour. The caudal fin is forked and a plain dull yellow, while the hind margin is dusky and the lower lobe is tipped white. This species is identifiable by the presence of two orange-yellow stripes that extend from the head to the caudal peduncle and the lack of dark bands on the caudal lobe.

Habitat: Coastal waters and estuaries, with a mud or silt substrate.

Diet: Feeds mostly on muddy substrates where it preys primarily on small crustaceans.

Range: Found throughout the Indo-Pacific. In Kuching, this fish is found between April and June in Sg. Kuap.

This small, rather attractive goatfish is often mistaken for Sillaginids when schooling together.

Yellow or Sulphur Goatfish (*Upeneus sulphureus*), *'Nangka'*.

EPHIPPIDAE
(Spadefish, Batfish and Scats)

Fishes of the family Ephippidae are found throughout the Atlantic, Indian, and Pacific oceans in marine and, occasionally, brackish waters. There are seven genera and 20 species in this family. Many books describe *Proteracanthus sarissophorus* as a member of the family Kyphosidae, commonly known as the Sea Chubs.

Sea Chub (*Proteracanthus sarissophorus*), '*Perencong*'.

Proteracanthus sarissophorus (Cantor, 1849)

Sea Chub

'Perencong'

Description: 35 cm. It is a very distinctive species with the fourth spine in the dorsal fin extending more than three times the length of the other dorsal spines and usually having an inverted hook at the tip. The body is deep and compressed laterally with a dull pinkish orange colour, darker on the upper body and slightly lighter on the lower sides and belly. The dorsal fin has 10 spines and 14 to 16 soft rays, while the anal fin has 3 distinctive spines and 13 to 15 soft rays.

Habitat: Marine, occasionally entering coastal waters and estuaries.

Diet: Omnivorous, feeding on algae and small invertebrates.

Range: Found throughout the Western Central Pacific, the Malay Peninsula, Sumatra and Borneo. In Kuching, it is an occasional visitor to Sg. Kuap.

MUGILIDAE

(Mullets)

Fishes of the family Mugilidae are found in most tropical and warm-temperate seas, with some found in cooler temperate waters. They mostly inhabit coastal and brackish waters, although some species occasionally enter freshwater, and a few species live almost entirely in freshwater. The body is elongated and sub-cylindrical while the head tends to be broad and depressed with a small to moderately sized mouth, which is terminal or inferior. There are two, well separated, dorsal fins; the first with spinous rays and the second with soft, segmented rays. The pelvic fins are sub-abdominal, with 1 spine and 5 soft rays. The lateral line is absent. The colouring tends to be silver, darker above and white below. Mullets usually travel in schools and adults feed on fine algae, diatoms and detritus of submerged objects and bottom sediments. The name Mugilidae is derived from the Latin 'mugil', referring to the fishes similar to mullets. The latest estimate has found 14 genera and 62 species in this family, although this number is likely to change with a review of the family.

Oedalechilus labiosus (Valenciennes, 1836)

Hornlip Mullet

'Gamura'* or *'Kemura'

Description: 40 cm, although usually caught in the waters around Kuching at lengths of 7 to 15 cm. The body is sub-cylindrical and slightly deeper than other mugilid species. The head is small, broad and depressed. The snout is relatively short and blunt with a thick-lipped terminal mouth and medium-sized eyes. The lachrymal (a dermal, lateral-line bone situated directly in front of the orbit, and also called preorbital, lachrimal, lacrimal) is distinctly notched. The body is silvery in colour, with an olive green tinge on the back and silvery white on the belly. There are usually 32 to 37 scales on the longitudinal series (the midline running from just behind the head and above the pectoral fin to the end of the caudal peduncle, not the caudal fin), and 11 or 12 scales in the transverse series (from the origin of the pelvic fins to the origin of the first dorsal fin). The first dorsal fin has 4 spinous rays and the second dorsal fin has 7 to 9 segmented rays in adults. The origin of the anal fin is usually anterior to the origin of the second dorsal fin, and has 3

Hornlip Mullet (*Oedalechilus labiosus*), *'Gamura'*.

spines and 9 segmented-rays. The fins are pale to slightly translucent and the caudal fin is emarginate (concave).

Habitat: Coastal waters, reef flats and shallow lagoon reefs.

Diet: Unknown, although can often be seen scraping the algae off the surfaces of floating and solid objects.

Range: Found throughout most of the Indo-Pacific. In Kuching, it is commonly found in Sg. Kuap, and is occasionally caught in Sg. Sarawak after Barrage activity.

Oedalechilus labiosus can often be seen schooling near the surface as they graze on algae-encrusted objects such as the supports of pontoons and jetties.

Liza tade (Forsskål, 1775)

Tade Mullet

'Belanak'

Description: 70 cm. It is usually caught in the rivers around Kuching at lengths of 20 to 35 cm. The body is elongated, slender and slightly sub-cylindrical, with approximately 30 to 35 scales in the longitudinal series and

Tade Mullet (*Liza tade*), *'Belanak'*.

ten or eleven scales in the transverse series. The head is small with a small, pointed, terminal mouth and medium sized eyes. The body colour is silvery overall, with a slightly darker greenish brown tint to the dorsal region and silvery white on the belly. The first dorsal fin has 4 to 5 spinous rays and the second dorsal fin has 9 segmented rays. The origin of the anal fin is usually anterior to the origin of the second dorsal fin and has 3 spines and 9 soft, segmented rays in adults. The dorsal fin is widely separated into the spinous and soft-rayed portions, with five spines and eight soft rays. The anal fin is slight anterior in the mirror image of the soft portion of the dorsal fin with three spines and nine soft rays. The caudal fin is truncate to emarginate (concave). This species may be mistaken for *Liza subviridis*, but the latter has fewer lateral line scales and the body is usually less elongate.

Habitat: Generally inhabits marine waters, but regularly enters estuaries and backwaters, such as lagoons, tidal creeks and also freshwater creeks, to feed.

Diet: Feeds on small algae, diatoms, benthic organisms and other organic matter, both living and detritus, sieved from the sand and mud.

Range: Found throughout much of the Indo-Pacific. In Kuching, it is found mostly in Sg. Kuap, but occasionally in the lower Sg. Sarawak.

The juveniles have been occasionally found in rice fields and mangrove swamps. Spawning takes place at sea.

POLYNEMIDAE
(Threadfins)

Fishes of the family Polynemidae are found in all tropical and subtropical seas. They generally swim in either pelagic and / or demersal waters. They are generally coastal inhabitants regularly entering estuaries and occasionally extending into the tidal zone of rivers. The name Polynemidae is derived from the Greek 'polys' meaning many and 'nema' or 'atos' meaning filament, translating as many filaments. This refers to the long filamentous free rays extending from the lower section of the pectoral fin. There are seven genera and 33 species in this family.

Leptomelanosoma indicum (Shaw, 1804)
Indian Threadfin
'Mancong'* or *'Senangin'

Description: 142 cm. Sometimes called *Polynemus indicum*, it is commonly caught between 50 and 80 cm. The body is moderately elongate and compressed with a sharply forked tail. The colour is grey, darker on the upper region and fading to a pale almost white belly. All the fins are darker. The head is relatively small and pointed. There are 5 pectoral filaments, the longest reaching to the base of the anal fin. Some smaller specimens appear to have filaments on the tips of the caudal forks. The spinous and soft-rayed parts of the dorsal fins are well separated with the anal fin opposite the second dorsal fin. The first dorsal fin has 9 spines, while the second has 13 to 14 soft rays. The anal fin has 2 to 3 spines and 11 to 12 soft rays.

Habitat: Marine, occurring mainly above shallow muddy and sandy bottoms of the continental shelf. It will occasionally enter rivers.

Diet: An active predator, it uses its filaments for feeding on benthic substrates, as well as hunting in mid and surface waters. The diet diversifies as the fish grows.

Range: Found throughout the Indo-Pacific. In Kuching, it occasionally enters Sg. Kuap.

Leptomelanosoma indicum is thought to enter inshore waters to spawn. A hermaphrodite species, it changes sex depending on size.

Indian Threadfin (*Leptomelanosoma indicum*). Smaller specimens are called *'Senangin'*, while *'Mancong'* is used for larger specimens.

Polynemus melanochir melanochir (Valenciennes, 1831)

Blackhand Paradise-fish

'Bulu'

Description: 25 cm. This striking fish is a beautiful metallic gold colour all over, slightly paler towards the belly. The pectoral fin has a black mark (hence the common name of 'Blackhand paradise fish') and 7 long black filament rays, the longest of which are almost twice as long as the body. The head is small with the mouth reduced and the eye small. The spinous and soft dorsal fins are well separated. The caudal fin is deeply forked. The soft dorsal fin and the anal fin are opposite one another and similar in size, appearing almost as mirror images.

Habitat: Marine, although may inhabit estuaries and the lower sections of rivers.

Diet: Unknown.

Range: Found throughout Asia. In Kuching, it commonly found in Sg. Kuap and Sg. Sarawak.

In Malay, '*bulu*' means hair; this refers to the long black 'hair-like' filaments.

Blackhand Paradise-fish (*Polynemus melanochir melanochir*), '*Bulu*'.

AMBASSIDAE
(Asiatic Glassfish)

Fishes of the family Ambassidae are found in marine, brackish and freshwater environments throughout Asia, Oceania, and the Indo-West Pacific oceans. As the name Glassfish suggests, many species of this family have a translucent or semi-transparent body, often with a silvery colouration. The fins are dusky or with pigmented blotches. They often occur in schools near river mouths while some species are reported to live in freshwater. The Ambassidae are commonly seen with a fluorescent coloured body in aquarium shops; this colouring is achieved by injecting a dye into the individual's body cavity. It is a cruel procedure from which fish usually do not survive for long.

This family was formerly known as Chandidae. The name Ambassidae is derived from the Greek 'ambasis' or 'anabasis' meaning climbing up. There are eight genera and 41 species in this family.

Ambassis nalua (Hamilton, 1822)
Scalloped Perchlet
'Perdin'* or *'Pridin'

Description: 12.5 cm. The body is a relatively deep oval shape and ventrally compressed, usually a silvery translucent colour. The head is small with large eyes and a medium terminal mouth with the lower jaw extending beyond the upper. The nasal spine is absent. The spinous and soft portions of the dorsal fin are separate, with 7 to 8 spines and 7 to 11 soft rays. The anal fin is the mirror image of the soft portion of the dorsal fin, having 3 spines and 7 to 11 soft rays. The pelvic fin has 1 spine and 5 soft rays, while the pectoral fin has usually 16 or 17 (rarely, 15) soft rays. The caudal fin is forked.

Habitat: Generally brackish water inhabitants of bays and estuaries, but regularly enter freshwater, where it can dwell in both running and still waters.

Diet: Unknown.

Range: Found throughout the Indo-West Pacific, from India to Papua New Guinea and Australia. In Kuching, it is found in Sg. Kuap and the more brackish waters of Sg. Sarawak.

Scalloped Perchlet (*Ambassis nalua*), *'Perdin'*.

Ambassis vachellii (Richardson, 1846)
Vachelli's Glass Perchlet
'Perdin'* or *'Pridin'

Description: 5.5 cm. The body is a slender oval shape and ventrally compressed, usually a silvery translucent colour. The head is small with medium to large eyes and a medium-sized terminal mouth with the lower jaw extending beyond the upper. The spinous and soft portions of the dorsal fin are separate, with 8 spines and 9 soft rays. The anal fin is the mirror image of the soft portion of the dorsal fin, having 3 spines and 9 soft rays. The pelvic fin has 1 spine and 5 soft rays, while the pectoral fin has usually 16 or 17 (rarely, 15) soft rays. The caudal fin is forked.

Habitat: Brackish waters of bays, estuaries and tidal mangrove creeks. It regularly enters freshwater, where it can dwell in both running and still waters.

Diet: Unknown.

Range: Found throughout the Indo-West Pacific. In Kuching, it is found in Sg. Kuap and Sg. Sarawak.

Vachelli's Glass Perchlet (*Ambassis vachellii*), *'Perdin'*.

DATNIOIDAE
(Tigerfish)

Fishes of the family Datnioidae are found from India to Borneo in fresh and brackish waters. The dorsal fin has 12 spines and 15 to 16 soft rays. The juveniles exhibit a remarkable defence mechanism against predators whereby they turn sideways and float, thus mimicking a dead leaf. The name of this family has been swapped in the past between Datnioidae and Coiidae and has recently reverted back to Datnioidae. There is only one genus and four species in this family.

Datnioides quadrifasciatus (Sevastianof, 1809)
Four-barred Tigerfish
'Rengau'

Description: 30 cm. The colour pattern of the body is highly variable between individuals, the numbers of vertical bars may be up to 7 full bars and between 1 and 4 partial bars. The caudal fin is rounded with a thick translucent white margin and a dark coloured base. The soft rays of the second dorsal fin are also a translucent white. The dorsal fin has 12 spines and 13 to 14 soft rays. The anal fin has 3 spines and 8 to 9 soft rays.

Habitat: Inhabits brackish water at the mouths of large rivers and coastal lagoons as well as in freshwater lakes and rivers above the tidal influence.

Diet: Carnivorous, feeding on fishes, prawns, crabs and some insect larvae.

Range: Found throughout Asia and Oceania from India to Indonesia and New Guinea. In Kuching, it is found in Sg. Kuap and the more brackish waters of Sg. Sarawak.

Datnioides quadrifasciatus is also a moderately common species in the aquarium trade.

Four-barred Tigerfish (*Datnioides quadrifasciatus*), *'Rengau'*.

APOGONIDAE
(Cardinalfish)

Apogonidae are found throughout the Atlantic, Indian, and Pacific oceans where they prefer marine habitats, but there are also brackish and some freshwater species. Most species are nocturnal, feeding on zooplankton and small benthic invertebrates. They are generally of small size with a maximum length of 20 cm although most species grow to less than 10 cm. This family is one of few that are mouth-brooders. After their eggs have been fertilised the male Cardinalfish scoops them into his mouth for several days of incubation, during which time no feeding takes place. The family is moderately popular with the aquarium trade.

Apogon hyalosoma (Bleeker, 1852)
Mangrove Humpbacked Cardinalfish
'Sebekah', 'Semangka', 'Seriding'?

Description: At 17 cm, *Apogon hyalosoma* is one of the larger species of Cardinalfish. It can easily be identified by a distinct black spot just in front

Mangrove Humpbacked Cardinalfish (*Apogon hyalosoma*), *'Sebekah'*.

of the caudal base about equal to eye diameter. The body colour is slight pinkish with a gold overtone. The trunk of the fish is squarish with the pre-dorsal area forming a light hump above the head. The eye is large, the snout blunted, with the maxillary reaching the rear border of the eye. The dorsal fins are separate, with the first dorsal fin having 7 spines and the second having 9 soft rays. The anal fin has 2 spines and 8 soft rays.

Habitat: Brackish water or the freshwater of river mouths.

Diet: Feeds on invertebrates.

Range: Found throughout Asia and Oceania, from Japan south to Indonesia and Papua New Guinea. In Kuching, it is found in Sg. Kuap.

Apogon hyalosoma is a nocturnal species.

THERAPONIDAE
(Grunters or Tigerperches)

Fishes of the family Theraponidae are found throughout the Indo-West Pacific in coastal waters, but there are several species restricted to freshwater. They generally feed on fishes, insects, algae, and sand-dwelling invertebrates. The name Theraponidae is also commonly spelled as "Teraponidae" or "Therapontidae" and is derived from the Greek 'tera', 'atos' meaning 'strange thing, monster, or wonder' and 'pontios' meaning marine, pertaining to the sea, translating as sea monster. There are 16 genera and 45 species is this family.

Terapon jarbua (Forsskål, 1775)
Crescent Perch or Jarbua Terapon
'Rumpak'* or *'Kerong-kerong'

Description: 36 cm. Commonly caught at lengths of around 15 to 20 cm. The body is oblong and slightly compressed with a silvery iridescent sheen, darker above and white below. It is easily identified by the 3 or 4 curved dark brown bands which run from the darkened nape region to the hind part of the body, the lowermost continuing across the middle of the caudal fin. The operculum bears 2 spines, with the lower extending well beyond the opercular flap. The ventral fins are a slight yellowish tan colour, the dorsal and caudal fins are transparent but with the continuation of dark bands. The dorsal fin is notched with 11 to 13 spines and 9 to 11 soft rays. The anal fin has 3 spines and 7 to 19 soft rays. The pelvic fin is situated behind the base of the pectorals and has 1 spine and 5 soft rays.

Habitat: Shallow sandy bottoms, in the environs of river mouths.

Diet: Omnivorous, feeding on fish, insects, algae and sand-dwelling invertebrates.

Range: Found throughout the Indo-Pacific. In Kuching, it is found in Sg. Kuap and the more brackish waters of Sg. Sarawak.

Adults of *Terapon jarbua* form loose aggregations, while juveniles are common in sandy inter-tidal areas such as tidal pools. It is often heard to produce a 'grunting' sound when pulled from the water.

Crescent Perch or Jarbua Terapon (*Terapon jarbua*), '*Rumpak*' or '*Kerong-kerong*'.

HAEMULIDAE
(Sweetlips and Grunters)

Fishes of the family Haemulidae are found in the Atlantic, Indian and Pacific oceans. They are mainly marine inhabitants, although there are some brackish water species which may occasionally enter freshwater environments. Adults are typically inactive during the day when they shelter near or under ledges, dispersing to feed on benthic invertebrates at night. They are pelagic spawners. The juveniles are popular in home aquaria, but adults require a large tank.

Their common name 'Grunters' comes from the sound they make when grinding their pharyngeal teeth, a sound amplified by their gas bladder. The name Haemulidae comes from the Greek 'haimaleos' meaning bloody. There are 17 genera and 150 species in this family.

Pomadasys kaakan (Cuvier, 1830)
Javelin Grunter
'Gerut-gerut'

Description: 80 cm, although usually caught between 15 and 50 cm. The body is slender, the head profile slightly convex, the mouth small with thickened lips and there are 2 pores and a central groove under the chin. The body colour is silvery overall with a golden tint on the sides. The upper operculum is olive-yellow and the snout dusky, overlain with yellow. In younger specimens there are 7 to 11 vertical bands on the upper sides of the body, while older specimens have bands formed by spots aligned in vertical rows. The dorsal fin is continuous, with a deep notch between the spinous and the soft-rayed portion. The fourth dorsal spine is the longest (the first is the shortest), with a total of 9 to 14 spines and 11 to 26 soft rays. The anal fin has 3 spines (the second of which is the longest) and 6 to 18 soft rays.

Habitat: Turbid inshore waters with sandy substrates. It will regularly enter estuaries as well as the lower reaches of tidally influenced rivers and may also be seen in association with inshore wrecks.

Diet: Feeds on crustaceans and fish.

Javelin Grunter (*Pomadasys kaakan*), *'Gerut-gerut'*.

RANGE: Found throughout the Indian Ocean and South China Sea, south to Queensland and north to Taiwan. In Kuching, it is found in Sg. Kuap.

LUTJANIDAE
(Snappers)

Fishes of the family Lutjanidae are found in tropical and subtropical areas of the Atlantic, Indian and Pacific oceans. They are mainly marine inhabitants, although they are known to enter estuaries and freshwater. Most of the species in this family are predators of crustaceans and fishes, while several are planktivores. Most thrive in aquaria conditions but tend to grow rapidly. They are generally demersal swimmers, extending down to depths of about 450 m. They are valued for their tasty flesh, but can sometimes lead to ciguatera (poisoning caused by eating the flesh of fish that is toxic to humans, in contrast to poisoning from eating old or putrefied flesh; this form of poisoning is unpredictable and sometimes fatal).

The name Lutjanidae is derived from the Malay name of the fish, '*ikan lutjan*'. There are 17 genera and 130 species in this family.

Lutjanus johnii (Bloch, 1792)
Red Mangrove Snapper
'Ikan Merah'

Description: 97 cm. The body is oval and compressed with a strong convex dorsal profile. The head is pointed, the eyes large, the mouth moderate to large and the jaws contain enlarged canine teeth. The body colour is yellow with a bronze to silvery sheen, shading to silvery white on the belly and the underside of the head. There is a distinctive large black blotch on and above the lateral line below the anterior dorsal fin rays although this fades slowly with age. The lateral line runs parallel to the dorsal profile. The centre of each scale often has a reddish brown spot giving an overall appearance of a series of horizontal lines along the body. The ventral fins are yellow with white margins, while the caudal and dorsal fins have a reddish tinge. The dorsal fin is continuous with a slight notch separating the spinous and soft-rayed portions, there being 10 dorsal spines and 13 to 14 soft rays. The anal fin has 3 spines and 7 to 11 soft rays. The pelvic fins originate just behind the pectoral base.

Red Mangrove Snapper (*Lutjanus johnii*), *'Ikan Merah'*.

Habitat: Adults are commonly seen around coral reefs, while the juveniles are found in mangrove estuaries.

Diet: Feeds on fish and benthic invertebrates including shrimps, crabs and cephalopods.

Range: Found throughout the Indo-Pacific region including the South China Sea. In Kuching, it is a regular visitor to Sg. Kuap.

GERREIDAE
(Mojarras)

Fishes of the family Gerreidae are found in the coastal waters of most tropical seas, with some species entering freshwater. Mojarras, also called biddies, are a family of silvery fish with compressed bodies and forked tails. They have highly protractile mouths which, when extended, point downwards. Gerreidae feed on benthic invertebrates, which they sort from sandy substrates.

The name Gerreidae is derived from the Latin 'gerres', meaning a kind of anchovies. There are eight genera and 40 species in this family.

Gerres oyena (Forsskål, 1775)
Common Silver Biddy
'Umba'

Description: 30 cm, although regularly caught between 5 and 12 cm. The dorsal fin has 9 spines and 10 soft rays; the first spine is the smallest and the second and third spines are the longest. The anal fin has 3 spines and 7 soft rays. Both the dorsal and anal fins have a sheath of scales along the base. There are 3 to 4 (usually 3.5) scales between the fifth dorsal spine and the lateral line. The lateral line is complete and runs parallel to the dorsal profile. The body is generally silvery in colour, but the dorsal region has a slightly greenish hue. The body has 5 to 6 indistinct dark vertical stripes, which may disappear in larger specimens. The ventral and anal fins are yellow with whitish tips while the pectoral fins are nearly colourless and transparent. The dorsal spine tips have an indistinct dusky patch.

Habitat: A coastal species and commonly found in estuarine areas.

Diet: Feeds on small organisms living on sandy substrates. Small juveniles feed on zooplankton, while larger fish feed on small polychaetes, bivalves, crustaceans and fish.

Range: Commonly found throughout the Indo-Pacific, Red Sea and India. In Kuching, is found in Sg. Kuap and the tidally influenced sections of Sg. Sarawak.

Gerres oyena is both a solitary and schooling species.

Common Silver Biddy (*Gerres oyena*), *'Umba'*.

LEIOGNATHIDAE
(Slimys, Slipmouths, or Ponyfish)

Fishes of the family Leiognathidae are found throughout the Indo-West Pacific. They are mainly a marine family, with the majority of species inhabiting the deeper waters where they feed on plankton, though many families are also common in shallow coastal waters and tidal creeks where they will eat small benthic invertebrates.

Ponyfish are distinctive as one of only eight shallow-water families, which are luminescent (most of the deepwater families have this ability to some extent). All the species in Leiognathidae have an oesophageal luminous organ, though whether all species use it is unclear. This luminescence is produced by symbiotic bacteria in the 'oesophageal diverticulum' (an organ in the gut). Light emission is controlled by the fish using 'musculature stutters' triggered by muscles located on the wall of the light organ. Patterns of light emission are distinctive to different species.

Leiognathidae are small fish with highly compressed bodies. Mouths are small and very protractile. As their common name 'slimy' suggests, they are also noted for their production of mucus. Leiognathidae comes from the Greek 'leio' meaning smooth or flat and 'gnathos' meaning jaw. There are three genera and 24 species in this family.

Leiognathus blochii (Valenciennes, 1835)

Two-blotch Ponyfish

'Kekek'

Description: 28 cm, although specimens are commonly found between 6 and 12 cm in length. The body is a deep oval shape, strongly compressed while the snout is blunt and the mouth, when protracted, points downward.

Two-blotch Ponyfish (*Leiognathus blochii*), *'Kekek'*.

The dorsal half of the body is silvery with numerous dark brown irregular markings. The ventral half of the body is silver. The fins are transparent with white spines.

Habitat: Thought to be coastal and estuarine, regularly entering rivers.

Diet: Unknown.

Range: In Kuching, found in both Sg. Sarawak and Sg. Kuap.

This species can be caught singly or in small schools.

Leiognathus equulus (Forsskål, 1775)
Common Ponyfish
'Kekek'

Description: 28 cm, although more commonly caught between 10 and 18 cm in length. The body is deep bodied and oval in shape, strongly compressed, while the snout is blunt and the mouth, when protracted, points

Common Ponyfish (*Leiognathus equulus*), *'Kekek'*.

downward. The dorsal fin comprises 8 spines and 15 to 16 soft rays. The first spine is diminutive, while the second is the longest. The anal fin has 3 spines and 14 to 15 soft rays. The soft rays in both the dorsal and anal fins have a straight profile. The body is silvery with the upper dorsal region more greyish with numerous dark vertical lines. The caudal peduncle has a small brown saddle. The anal fins are yellowish, while the dorsal fins are transparent.

HABITAT: River mouths and muddy inshore areas such as mangroves. It frequently ascends into the freshwater reaches of rivers.

DIET: Feeds on polychaetes, small crustaceans, small fishes and worms.

RANGE: Found throughout the Indo-West Pacific. In Kuching, it is found mostly in Sg. Kuap, but occasionally in the estuarine reaches of Sg. Sarawak after Barrage activity.

Leiognathus equulus is diurnal. The adults move in schools.

Secutor ruconius (Hamilton, 1822)

Deep-pugnose Ponyfish

'Kekek'

DESCRIPTION: 8 cm. A small fish with a deep and strongly compressed body and the mouth is small and points upwards when protracted. The lateral line runs parallel to the dorsal profile but ends just before the caudal peduncle (the middle of the soft rays of the dorsal fin). The dorsal fin has 8 spines (the first of which is diminutive, while the second is the longest) and 16 soft rays. The anal fin has 3 spines and 14 soft rays. The body is a dark silvery colour above with a series of indistinct green or blue blotches forming stripes, which become progressively smaller towards the tail. The lower half of the body is silvery white in colour. The caudal fin has a slight yellow tinge while all the other fins are colourless and translucent. There is a slight black band extending from the lower margin of the eye to the angle of the lower jaw.

HABITAT: Inhabits the warm, shallow coastal waters where it enters estuaries and rivers.

DIET: Feeds mainly on small crustaceans.

RANGE: Found in the Indo-West Pacific, Indian Ocean and throughout South-East Asia. In Kuching, it is found in Sg. Kuap.

Secutor ruconius is a demersal schooling species.

Deep-pugnose Ponyfish (*Secutor ruconius*), *'Kekek'*.

SCIANEIDAE
(Drums or Croakers)

Fishes of the family Scianeidae are found in the Atlantic, Indian and Pacific oceans. The dorsal fin is long, with a deep notch between the spinous and soft-rayed parts; however, these parts are rarely completely separate. The swim bladder, often the best way of identifying the species, generally has many branches and is used as a resonating chamber. Scianeidae have exceptionally large otoliths. They are bottom-dwelling carnivores, feeding on benthic invertebrates and small fishes.

The name Scianeidae comes from the Greek 'skiaina', which is a kind of fish. The common name 'Croaker' comes from the croaking sound they produce by contracting their swim bladders. The sound is made not only during distress, but also for other purposes, and can be heard by divers underwater. There are 70 genera and 270 species in this family.

Otolithoides biauritus (Cantor, 1849)

Bronze Croaker

'Paloh'

Description: 160 cm, although commonly caught between 50 and 80 cm. The body is elongate with a large blunt head and hunched back tapering to a relatively narrow caudal peduncle. The lateral line is almost parallel with the dorsal profile. The body colour is dark brass and the fins dark brown. The dorsal fin extends from the shoulder to the caudal peduncle with 9 to 10 spines in the pre-notch portion and 1 spine and 27 to 32 soft rays in the post-notch portion. The anal fin has 2 weak spines and 6 to 7 soft rays. The caudal fin is large and rounded.

Habitat: Unknown.

Diet: Feeds on small fish and invertebrates.

Range: Found in coastal waters throughout the Indo-West Pacific. In Kuching, it is found in Sg. Kuap.

Bronze Croaker (*Otolithoides biauritus*), *'Paloh'*.

Nibea soldado (Lacepède, 1802)
Soldier Croaker
'Gelama batu'

Description: 60 cm, although usually caught between 20 and 40 cm. The body is oval in shape, elongate and compressed. The head is medium with a blunt snout and moderate mouth containing both large and small teeth. On the chin the first pair of mental pores are close together and united by a crescent-shaped groove. The body is silver in colour, a darker brownish-green above and silvery-white below. The dorsal fins are clear and transparent, while the ventral fins have a yellow tinge. The dorsal fin has 10 to 11 spines in the pre-notch portion and 27 to 33 soft rays in the post-notch portion. The anal fin has 2 spines, the second being long and stiff, and 7 soft rays. There are 8 to 13 scale rows above the lateral line to the origin of the dorsal fin and 12 to 17 scale rows below the lateral line to the origin of the anal fin.

Identification can also be confirmed by dissection, where the swim bladder is carrot-shaped, with 18 to 22 pairs of arborescent (tree-like) appendages along its sides, the first pair entering the head beyond the transverse septum. The sagitta (largest of the otoliths or ear-stone) has a tadpole-shaped impression, the tail of which is J-shaped.

Habitat: Shallow coastal and estuarine waters. Juveniles occur in brackish estuaries and often ascend the lower reaches of large, turbid rivers.

Soldier Croaker (*Nibea soldado*), *'Gelama batu'*.

Diet: Feeds on small fishes and invertebrates.

Range: Found throughout the Indo-West Pacific. In Kuching, it is found in Sg. Kuap and the more brackish waters of Sg. Sarawak.

Panna microdon (Bleeker, 1849)

Panna Croaker

'Gelama pisang'

Description: 30 cm. The body is elongate and slender, with a large head, relatively medium to small eye and a large mouth containing both large and small teeth. The body is a brownish-grey above and silver on the flanks and belly. The fins are yellowish with a darker margin on the dorsal and anal fins. The dorsal fin has 10 to 11 spines in the pre-notch portion and 33 to 36 soft rays in the post-notch portion. The anal fin has 2 spines, the second being weak, and 6 to 7 soft rays. The lateral line curves from the shoulder to the middle of the caudal peduncle. The caudal fin is rhomboid or diamond shaped.

Identification can also be confirmed by dissection; the swim bladder is carrot-shaped and enters the head. The sagitta (largest of the otoliths or ear-stone) has a tadpole-shaped impression, the tail of which has a sharp bend.

Habitat: Shallow coastal and estuarine waters. Juveniles are common in mangrove swamps.

Diet: Feeds on prawns, crustaceans and small fish.

Panna Croaker (*Panna microdon*), *'Gelama pisang'*.

Range: Found throughout the Western Central Pacific. In Kuching, it is found in Sg. Kuap and the more brackish waters of Sg. Sarawak.

Bahaba polykladiskos (Bleeker, 1852)
Spine Bahaba
'Turusan'

Description: 40 cm. The body is elongate with a large head and a steeply curved dorsal profile. The body colour is brownish, fading to almost white below. The fins are all a darker brown with the pectoral fin a translucent brown. The pre-notch portion of the dorsal fin has 11 spines while the post-notch portion has 27 to 30 soft rays. The anal fin has 2 spines and 7 soft rays.

Habitat: Shallow coastal and estuarine waters.

Diet: Unknown.

Range: Found throughout the Western Pacific, Viet Nam, Cambodia, and Borneo. In Kuching, it is found in Sg. Kuap and the more brackish waters of Sg. Sarawak.

Spine Bahaba (*Bahaba polykladiskos*), *'Turusan'*.

TOXOTIDAE
(Archerfish)

Fishes of the family Toxotidae are found from India to the Philippines, Australia and Polynesia in coastal, brackish and freshwater habitats where they are commonly seen near the water surface. They make use of the unique hunting technique of spitting water at prey (see frontispiece), which consists of mainly terrestrial insects, thereby knocking them into the water. Archerfish have large complex eyes, possibly binocular, enabling them to accurately discern distance and optic distortion in order to strike a target at up to 1.5 m distance. The name Toxotidae comes from the Greek 'toxotes' meaning archer or bowman. There is one genus and six species in this family.

Toxotes chatareus (Hamilton, 1822)
Largescale Archerfish
'Sumpit'

Description: At 40 cm, *Toxotes chatareus* is the largest species of Toxotidae. The body is compressed and deep-bodied with a straight back and deeply curved belly. The dorsal fin is situated posteriorly towards the tail and has 4 to 6 strong spines and 11 to 14 soft rays. The anal fin has 3 spines and 15 to 18 soft rays. There are approximately 25 to 47 scales on the lateral line. The body is silvery in colour with alternating large and medium black bands along the dorsal region. These black bands usually extend to cover the caudal peduncle, while the caudal fin itself remains pale in colour. There are black markings on the soft-rayed portion of the dorsal fin and the anal fin can be completely dark. The number, size and sharpness of the black bands vary among individuals. There can be difficulty differentiating between smaller specimens of *T. chatareus* and *T. jaculatrix*, but generally *T. jaculatrix* has even black bands rather than alternate large and small black blotches on the dorsal area of the body. Also, *T. chatareus* has a golden rim above the eye.

Habitat: Coastal areas, mainly in brackish mangrove estuaries, although it has also been spotted under the jetties of coral reef island resorts. It is also found in freshwater rivers and small streams, where it prefers shaded areas with overhanging vegetation.

DIET: Carnivorous, feeding on terrestrial insects either by jumping out to catch them or more famously spitting jets of water to knock them into the water. Also feeds on small fishes, prawns and insect larvae.

RANGE: Found throughout warm waters of the Indo-Pacific. In Kuching, it is found in Sg. Kuap and Sg. Sarawak.

Largescale Archerfish (*Toxotes chatareus*), '*Sumpit*'.

SCATOPHAGIDAE
(Scats)

Fishes of the family Scatophagidae are found throughout the tropical areas of the Indo-Pacific where they inhabit coastal regions often around harbours, estuaries and mangroves, even frequently entering freshwater. They have deep, angular and strongly compressed bodies similar to that of the butterflyfish.

The name Scatophagus has been given due to their habit of eating refuse or offal, and can be translated as 'excrement-eater', derived from the Greek 'skatos' meaning excrement and 'phagein' meaning to eat. The family comprises two genera and four species, of which only *Scatophagus argus* is regularly found in Malaysian waters. They are marketed for human consumption but their feeding habits can be off-putting. They are also commonly found in the aquarium trade as active and curious little fish when juvenile.

Scatophagus argus (Linnaeus, 1766)
Spotted Butterfish or Spotted Scat
'Kitang'

DESCRIPTION: 38 cm. A deep-bodied and strongly compressed squarish fish. The body and head are covered in minute adhesive scales and the lateral line is arched continuously to the anterior caudal peduncle region. The head is small and triangular in shape with a slight concave angle just above the

Spotted Butterfish or Scat (*Scatophagus argus*), *'Kitang'*.

eye. The dorsal fins are clearly separated into spinous and soft-rayed sections with 10 to 11 spines and 16 to 18 rays. The dorsal spines are thought to be poisonous. The anal fin has 4 spines and 13 to 15 rays. The soft rays of the dorsal and anal fins give a square appearance to the body. The caudal fin is truncated or slightly rounded. The body is yellowish to brown in colour, covered with dark brown or black spots. In some specimens the spots are barely distinguishable from the dark body, while in juvenile specimens the spots are brighter. Juveniles of 1 cm length have been observed to have a red spot on the front of the head. These spots may fade from the lower body as the fish grows.

Habitat: Marine to freshwater environments.

Diet: Highly omnivorous and opportunistic feeders, from aquatic vegetation, small prawns and fish to offal and waste. Juveniles feed mainly on algae and macro invertebrates. As they grow their diet expands to include small fish and shrimps.

Range: Found throughout the Indo-Pacific. In Kuching, it is found in Sg. Kuap and the more brackish waters of Sg. Sarawak.

Juveniles of *Scatophagus argus* group together and are often found up-river in small tributaries in shallow waters where they feed mainly on algae.

CARANGIDAE
(Scads and Trevallies)

Fishes of the family Carangidae are found throughout the Atlantic, Indian and Pacific oceans. They are mostly marine inhabitants, although occasionally found in brackish waters. The body can vary in size, but the spinous and soft-rayed portions of both the dorsal and anal fins remain separate. The caudal peduncle is slender and the caudal fin forked. The anterior section of the lateral line is arched, while the posterior section is straight.

Fish of this family are all fast-swimming predators preferring the open sea and the waters above reefs. One of the most important families of tropical marine fishes, they are highly valued in both commercial and sports fisheries. The name Carangidae originates from the French 'carangue' which was the name given to one of the species in Caribe. There are 33 genera and 140 species in this family.

Most members of this family that enter the estuarine waters around Kuching are juveniles which often have not yet obtained their adult colouring. This makes them difficult to identify.

Scomberoides commersonnianus (Lacepède, 1801)
Juvenile Talang Queenfish

Description: 120 cm. The body is compressed and an elongate oval shape. The eye is large and the maxilla extends beyond the eye. Soft dorsal and anal fin origins are opposite and diagonally positioned. The dorsal fin has 7 to 8 spines and 19 to 21 soft rays. The anal fin has 3 spines; the first 2 spines are separate from the rest of the fin, which has 16 to 19 soft rays. The colour is a dusky green above and silvery below, while the dorsal lobe is dusky to dark and evenly pigmented. In adults, there are 6 to 8 oval spots along the body just above the lateral line. Juveniles tend to be a solid silver colour with no distinct markings.

Habitat: Coastal waters. It is a regular visitor to reefs and offshore islands. It is also an occasional visitor to estuaries.

Juvenile colouration of the Talang Queenfish (*Scomberoides commersonnianus*).

DIET: Feeds on fishes, cephalopods, and other pelagic prey.

RANGE: Found throughout the tropical waters of the Indo-West Pacific. In Kuching, juveniles can be found in Sg. Kuap.

Scomberoides commersonnianus usually congregates in small groups.

Caranx sexfasciatus (Quoy & Gaimard, 1825)

Juvenile Bigeye Trevally

'Belekok'

DESCRIPTION: 120 cm. The body is compressed and slightly elongate. The mouth is moderately large with a prominent lower jaw. The dorsal fin has 9 spines and 19 to 22 soft rays. The anal fin has 3 spines, two of which are detached, and 14 to 17 soft rays. The body colour is iridescent blue-green dorsally and silvery white below, while the soft dorsal fin and the anal fin have white-tipped lobes. The pectoral fins are sickle shaped and yellowish to black in colour. The upper region of the opercle has a small blackish spot.

Juvenile colouration of the Bigeye Trevally (*Caranx sexfasciatus*), *'Belekok'*.

The anterior portion of the lateral line is curved, while the posterior portion is straight. The straight portion has 27 to 36 strong, dark scutes. Juveniles are generally silver in colour with 6 vertical faint dusky bands along the body.

Habitat: Coastal and oceanic waters. It is commonly associated with reefs. Juveniles are commonly found in estuaries and occasionally enter rivers, possibly penetrating well inland.

Diet: Feeds mostly on fish and crustaceans.

Range: Found throughout the Indo-Pacific. In Kuching, the juveniles can be found in Sg. Kuap.

Caranx sexfasciatus forms slow-moving schools in reef passes or outside reefs during the daytime and disperse at night to hunt.

SIGANIDAE
(Rabbitfish)

Fishes of the family Siganidae are found throughout the tropical waters of the Indo-Pacific and eastern Mediterranean. They are mostly pelagic marine inhabitants, but are known to enter estuaries. The pelvic fins have 3 soft rays between an inner and an outer spine. The dorsal fin has 13 strong spines and 10 soft rays, while the anal fin has 7 spines and 9 soft rays. The spines are poisonous.

Some species are schooling, while others live among the coral structures, generally seen in pairs. All species of this family are diurnal and herbivores, feeding on benthic algae. Some of the more colourful species are popular in the aquarium trade. The name Siganidae is derived from the Latin 'siganus' meaning rabbit, referring to the similarity of the nose shape. There are only two genera and 25 species in this family.

Siganus javus (Linnaeus, 1766)
Juvenile Streaked Spinefoot/ Rabbitfish
'Debam'* or *'Dengkis'* , *'Ketang'?

Description: 53 cm. The body is oval and strongly compressed. The nape is slightly concave and the rest of the head is convex. The mouth is small and terminal, with fleshy 'rabbit-like' lips. In juveniles there is a broad-based triangular flap of anterior nostril reaching half-way to the posterior nostril, reducing to a small crescent shape with age. The lateral line runs almost parallel to the dorsal profile. The dorsal fin has 13 slender and thorn-like sharp spines and 10 soft rays. The anal fin has 7 stout spines and 9 soft rays, while the pectoral fin has 2 spines separated by 3 soft rays. The body is bronze in colour above, fading to white on the belly and thorax, with numerous pale-grey blotches that lengthen and merge into long wavy lines along the ventral region. Juveniles develop very distinct blotches when alarmed. The pectoral fins are transparent and the pelvic fins are white. There is no orange blotch above the caudal peduncle in this species.

HABITAT: Generally associated with the shallower waters of reefs, although it is also found in the brackish water of estuaries.

DIET: Feeds on algae attached to the substrate and on floating algal fragments.

The mottled colouration of a juvenile Streaked Spinefoot or Rabbitfish (*Siganus javus*) displayed when alarmed.

Range: Found in tropical waters of the Indo-Pacific. In Kuching, juveniles can be found in the mangrove areas of Sg. Kuap.

Adult *Siganus javus* school together in small groups of 2 to 10 individuals. They can often be seen floating in mid-water at depths of 2 to 6 m when not feeding. Important to both commercial fisheries and the aquarium trade.

GOBIIDAE
(Gobies)

Fishes of the family Gobiidae are found mostly in tropical and subtropical waters. They are found in shallow coastal waters, reefs and freshwater habitats with some species being catadromous, meaning they migrate from freshwater to the sea to spawn. The body is usually cylindrical and compressed, with a round or compressed head. They have 2 dorsal fins, the first generally having 6 spines and the second having 1 spine and 5 to 37 soft rays. The anal fin has 1 spine and 5 to 36 soft rays. The pelvic fins vary between species, with degrees of separation from being completely separate to being completely fused into a disc-like form, the latter being the more typical. Most Gobiidae have an arrangement of sensory papillae on the head, the arrangement of which is often used in species identification. The caudal fin is either pointed or rounded.

The Gobiidae are a family of small fish growing to a maximum length of 50 cm, while most species are below 10 cm in length. The Gobiidae is the largest family of marine fishes (possibly comprising more than 2000 species) and includes the smallest fishes in the world. Most of the Gobiidae are cryptic bottom-dwelling carnivores feeding on small benthic invertebrates, while others are planktivores. Some species have symbiotic relationships with invertebrates (e.g., shrimps) and others are known to remove ecto-parasites from other fishes. They are nest spawners with non-spherical eggs guarded by the male. Some are popular aquarium fishes. The following subfamilies are recognized: Oxudercinae, Amblyopinae, Sicydiinae, Gobionellinae and Gobiinae.

There appears to very little readily available literature on the Gobiidae family and this has made identification difficult. A new book entitled *A Guide to the Gobies of Singapore* by Dr Helen Larson and Dr Kelvin Lim Kok Peng published by Singapore Science Centre will help to address this lack of literature.

Brachygobius doriae (Günther, 1868)
Bumblebee Goby
'Gendot'

Description: 4.2 cm. A small goby; the body resembles that of a bumblebee with four black and yellow bands, hence the common name 'bumblebee goby'. The first black band on the body starts at the nape and ends at the base on the first dorsal fin. The second black band starts by crossing the middle of the dorsal and anal fins and ends at the start of the caudal peduncle, while the third black band starts at the middle of the caudal peduncle and extends to the caudal base. Each black band is wider than the posterior yellow band. The pectoral and ventral fins are black for the first half to three-quarters of the fin while the remainder is a translucent white. The caudal fin is yellowish.

Habitat: Lives in various habitats from clear forest streams to tidal creeks and man-made canals.

Diet: Feeds on small fish, insects and mosquito larvae.

Range: Found through most of Asia, including Indonesia, Malaysia, Brunei and Singapore. In Kuching, it is commonly found amongst the aquatic plants along the banks of Sg. Sarawak.

Brachygobius doriae is very popular among the aquarium trade in Europe.

Bumblebee Goby (*Brachygobius doriae*), *'Gendot'*.

Glossogobius sp.
'Bekut'

Description: There are four species of *Glossogobius* that may be present in the rivers around Kuching: *G. aureus*, *G. celebius*, *G. circumspectus* and *G. giuris*. One main distinguishing feature between these four concerns the numbers and arrangement of the sensory papillae on the side of the head. All specimens were found between 6 and 30 cm in length. The bodies are all elongate with relatively long and depressed heads. The mouths are large with fleshy lips, the lower jaw longer than the upper. The colour ranges from pale cream to dark brown. The fins are all translucent with the dorsal and caudal fins having cross-sectional or diagonal narrow pale bands. There is usually a dark band from the lower eye margin to the upper jaw. Most specimens have five blotches or spots running from behind the pelvic fin to the base of the caudal. These blotches vary in definition, clarity and brightness, with some specimens appearing to have five distinct spots while others have a pale body and one distinct black spot on the caudal base. Longitudinal bands, which run along the upper and lower margins of the five blotches, are also a variable feature. Some are distinct, while others are faint or absent entirely. Many of the specimens also appear to have iridescent patches on the gill covers. The final difference between these specimens is the head length. Some specimens have long and depressed heads, while others have shorter, blunter depressed heads.

Habitat: Species of *Glossogobius* are generally estuarine but often found in freshwater and coastal waters, with the exception of *G. aureus,* which is a freshwater species.

Diet: Many of the specimens caught had a half-swallowed prawn in their throat; small fish and crustaceans may also be apart of their diet.

Range: Most of the species of *Glossogobius* are found throughout the Indo-Pacific. Specimens that had many of the different characteristics described above were caught from both the upper and lower tidal zones of the Kuap and Sarawak Rivers in Kuching.

(Opposite). The various colourations of three species of the genus *Glossogobius*, *'Bekut'*.

Acentrogobius viridipunctatus (Valenciennes, 1837)
Spotted Green Goby
'Bekut'

Description: Thought to reach a maximum standard length of only 10 cm, but specimens caught in the rivers around Kuching are approximately 15 cm in length. The body is moderately elongated with a short blunt head and a terminal mouth. The upper part of the eyes is slightly raised above the head surface. The body is a light brown colour, lighter on the belly, with electric blue and green coloured spots that form 5 to 7 irregular circles along the length of the body. The spots on the dorsal part of the circles are electric blue and those on the ventral part are electric green. The fins are transparent, with the pectoral and pelvic fins having a narrow white margin. The caudal fin is rounded and has a slight reddish coloured margin.

Habitat: Coastal waters and the brackish tidal waters of estuaries and rivers.

Diet: Unknown.

Range: Found throughout the Indo-West Pacific from east Africa to Japan and eastward to Papua New Guinea. In Kuching, it is found in Sg. Kuap and the more brackish waters of Sg. Sarawak.

Spotted Green Goby (*Acentrogobius viridipunctatus*), *'Bekut'*.

Ophiocara porocephala (Valenciennes, 1837)
Northern Mud Gudgeon
'Bekut'

Description: 34 cm. The body is relatively slender with a broad and slightly pointed snout. The eyes are relatively large. The body is brownish-pinkish in colour, lighter on the ventral region. Many of the scales have a lighter centre, giving the impression of numerous lines along the body. The body has larger areas of darker and lighter patches, and most specimens show two lighter bands or saddles, the first just posterior to the second dorsal fin and the second at the posterior end of the dorsal fin, just prior to the caudal peduncle. These pale saddles may be more prominent in juveniles, fading in larger specimens. The fins are transparent, with the dorsal and anal fins patterned with rows of pale dots. The dorsal, anal and caudal fins all have narrow, yellow margins. The dorsal fin has 7 spines and 8 to 9 soft rays. The anal fin has 1 spine and 7 soft rays.

Habitat: Brackish waters, estuaries, river mouths and freshwater rivers.

Diet: Unknown.

Northern Mud Gudgeon (*Ophiocara porocephala*), *'Bekut'*

Range: found throughout the Indo-West Pacific. In Kuching, it is found in Sg. Kuap and the more brackish waters of Sg. Sarawak.

Stenogobius sp.
'Bekut'

Description: Specimens were caught at around 10 to 13 cm in length. The body is elongated, with the head being rounded and blunt. The eyes are relatively large and the body is a light golden brown, fading towards the ventral regions, with two slightly lighter bands running the entire length of the body. There is a dark spot behind the gills, at the base of the pectoral fins. The fins are translucent, with the top half of the dorsal and caudal fins having a faint checkered black and white pattern. The first dorsal fin and the posterior spines appear to have filaments that are approximately 1 to 1.5 cm long. The caudal fin is fan shaped and rounded.

Habitat: Unknown.

Diet: Unknown.

Range: In Kuching, found in both the Kuap and Sarawak rivers.

Stenogobius sp.

Oligolepis **sp.**

'Bekut'

Description: This *Oligolepis* sp. measures around 7 to 12 cm in length. The body is elongate with a rounded and blunt head. The eyes are relatively large and situated towards the top of the head. The body is pale pink in colour with translucent fins, and the dorsal and caudal fins have an orange or brown series of dots. The base of the caudal fin has a central black spot.

Habitat: Thought to be estuarine.

Diet: Unknown.

Range: In Kuching, *Oligolepis* sp. is found in Sg. Kuap and the more brackish waters of Sg. Sarawak.

Oligolepis sp.

Mudskippers are amphibious air-breathing gobies, are currently classified in the subfamily Oxudercinae, but some older books and references have classified them in their own family: Periophthalmidae. They are associated with mangroves and tidal mudflats in tropical and subtropical waters. They inhabit burrows in the mud, which they defend rigorously. They are the only group of fish to live a considerable proportion of their life on land, including feeding and courting behaviours. Adaptation to living out of water has led to a number of physiological and morphological changes such as the eyes, which have adapted for clear vision both in and out of water; the fins, which have adapted to help them walk, climb and leap over land as well as swim, and the skin and gill chambers, which have become specialised by enhancing oxygen exchange to allow them to breathe both in water and air.

Serpent Mudskipper (*Parapocryptes serperaster*), *'Illung'*.

Parapocryptes serperaster (Richardson, 1846)

Serpent Mudskipper

'Illung'

Description: 23 cm. The body is elongate with a small rounded and blunt head. The eyes are located on the top of the head and are raised above the head surface. The body is light brown in colour with approximately five darker brown spots running along the flank. These are joined to posterior spots that run along the dorsal area. The fins are translucent, with the ventral edges of the pectoral and caudal fins having a narrow peach-coloured margin. The caudal fin is long, and fan shaped.

Habitat: Generally inhabits bays and estuaries but is also found in the freshwater tidal zones of rivers.

Diet: Feeds on fine algae and detritus.

Range: Found throughout the Indo-West Pacific including India, Indo-China, China, Malaysia and Indonesia. In Kuching, it is found in Sg. Kuap and the more brackish waters of Sg. Sarawak.

Periophthalmodon schlosseri (Pallas, 1770)

Giant Mudskipper

'Tamakul'* or *'Tembakul'

Description: At 27 cm, *Periophthalmodon schlosseri*, as the common name 'Giant Mudskipper' suggests, is one of the largest mudskippers. The body is elongate with a large blunt head, which has a pug-like appearance, with large teeth and rather fleshy lips. The eyes are large and protrude above the head. The dorsal area of the body is a dark brown colour, the sides are a lighter brown with numerous pale blue spots, and the belly is pale.

Habitat: Commonly inhabits the mudflats of estuaries, mangroves and the tidal zone of rivers. This species lives in deep burrows in the mud and emerges at low tide.

Diet: Thought to graze on the algae growing on the mudflats. Has been observed to hunt and prey upon other species of mudskippers.

Range: Found throughout the Indo-West Pacific. In Kuching, it is found across the mudflats of Sg. Kuap and the more brackish areas of Sg. Sarawak.

Periophthalmodon schlosseri can move very quickly across the mudflats when threatened, diving either into the burrow or into the water.

Giant Mudskipper (*Periophthalmodon schlosseri*), *'Tamakul'*.

Boleophthalmus boddarti (Pallas, 1770)
Blue-finned Mudskipper
'Layar'

Description: 22 cm. The body is elongate with a large head that is less blunt than *Periophthalmodon schlosseri*, with a large mouth. The eyes are large and protrude above the head. The entire body is brown in colour, but lighter on the belly. Dark brown marks on the head, dorsal region and flanks give the appearance of indistinct diagonal bands. The body, especially the fins, is covered in both medium and small blue spots. The first dorsal fin, when extended, is brightly coloured with five long blue streamers.

Habitat: Inter-tidal mudflats, where they can be seen raising and lowering their brightly coloured fins in territorial displays. During high tide, these mudskippers lie concealed in air pockets in their mud burrows, waiting to emerge at low tide.

Blue-finned Mudskipper (*Boleophthalmus boddarti*), *'Layar'*.

Blue-finned Mudskipper (*Boleophthalmus boddarti*), *'Layar'*.

Diet: They can be seen sweeping their heads from side to side as they feed on the marine algae covering the mud surface.

Range: Indo-West Pacific, from India to New Guinea and north to China. In Kuching, it is found across the mudflats of Sg. Kuap and the more brackish areas of Sg. Sarawak.

When a rival comes too close, a fierce battle may follow, mouths gaping wide and dorsal fins erect. A brief showdown takes place, and the loser retreats to its own burrow.

ELEOTRIDAE
(Sleepers)

Fishes of the family Eleotridae are found throughout most tropical and subtropical fresh waters. They have separate pelvic fins. The mouth is terminal. The dorsal fin has 2 to 8 flexible spines. The name Eleotridae comes from the Greek 'eleotris', which is the name of a particular Nile fish. There are 35 genera and 150 species in this family.

Butis amboinensis (Bleeker, 1853)
Olive Flathead-gudgeon
'Bekut'* or *'Puntang'?

DESCRIPTION: 14 cm. The body is elongate, with a depressed and relatively long head having a broad dark stripe roughly equal in width to that of the pupil, running from the eye to the upper jaw. The body is a dark brown colour with small reddish spots associated with the scales. There is a clear red, orange or peach coloured spot on the base of the pectoral fins. The first

Olive Flathead-gudgeon (*Butis amboinensis*), *'Bekut'*.

dorsal fin has 7 spines and the second has 8 soft rays. The anal fin has 1 spine and 8 soft rays.

Habitat: Generally inhabits brackish, mangrove waters, creeks and rivers, sometimes well upstream.

Diet: Unknown.

Range: Found throughout the Indo-Pacific, In Kuching, it is found in Sg. Kuap and to a lesser extent in Sg. Sarawak.

There is another very similar species, *B. humeralis*, which may also be present in the region. The difference between the two species is that the interorbital space (between the eyes) is densely scaled in *B. humeralis,* but almost naked in *B. amboinensis*, and the interopercle (middle bone of the operculum) has at least one row of scales in *B. humeralis,* but is naked in *B. amboinensis.*

Oxyeleotris marmorata (Bleeker, 1852)
Marble Goby
'Betutu'

Description: At 65 cm, *Oxyeleotris marmorata* is the largest species of the family Eleotridae. The body is elongate with a blunt, rounded head; the eyes are relatively small and situated towards the top of the head. The mouth is large and is the same width as between the eyes. The body is generally a dark brown colour with paler almost yellow markings; one of these markings is a distinctive 'arrow-head' at the caudal peduncle that points forward along the body. The other pale bands are not so clear in shape. The ventral region is paler with an off-white belly. The first dorsal fin has 7 spines and the second has 9 soft rays. The anal fin has 1 spine and 8 soft rays.

Habitat: Rivers, swamps, reservoirs and canals and will enter flooded forests.

Diet: Feeds on small fish, shrimps, aquatic insects, molluscs and crabs.

Range: Found throughout South-East Asian waters, from Malaysia, Indo-China, the Philippines and Indonesia. In Kuching, it is found in Sg. Sarawak.

Oxyeleotris marmorata is considered a delicacy over much of eastern Asia, and hence can be exported at high prices. It is commonly found in aquaria in the restaurants throughout Sarawak.

Marble Goby (*Oxyeleotris marmorata*), *'Betutu'*.

Oxyeleotris urophthalmoides (Bleeker, 1853)
'Bekut'

Description: 19.8 cm. The body is elongate with a blunt, depressed head. The eyes are situated towards the top of the head and there are two broad dark stripes, roughly equal in width to that of the pupil, running from the upper jaw to each eye and continuing past the eyes, fading at the operculum. The body is brown in colour with each scale having a paler spot, giving an impression of numerous narrow lighter lines running along the body. The top of the caudal base has a large black spot, which is similar to the identifying character of the gudgeon *Bostrichthys sinensis,* but easily separated by its cycloid scales. *O. urophthalmoides* and *O. urophthalmus* are very similar species and can be differentiated because *O. urophthalmoides* has 60 to 70 lateral line scales, while *O. urophthalmus* has 85 to 102.

Oxyeleotris urophthalmoides, 'Bekut'.

Habitat: Freshwater inhabitant extending its domain downriver to the top of the tidal zone.

Diet: Unknown.

Range: Found throughout South-East Asia, mostly in Indonesia. In Kuching, it is found in the fresher waters of Sg. Sarawak.

Pogoneleotris heterolepis (Günther, 1869)
'Batu Betutu'

Description: There is very little information on the characteristics or biology of this species. The specimens found in the rivers around Kuching were fairly large, reaching approximately 30 cm. The bodies were elongate but with a rectangular, thick body narrowing around the caudal peduncle to a long, rounded tail. The head is blunt and covered with short hair-like papillae and the eyes are black and very small, situated near the top of the head. All specimens from the Kuching area were dark brown to almost black in body colour. This colouring may be unusual as the few other known specimens are apparently lighter with a pink colouring.

Habitat: Brackish waters, where it is thought to be demersal. Fishermen say that it is found on muddy sediments and usually associated with muddy waters, hence the reduced eyes.

Diet: Unknown.

Range: May be found throughout the Indo-Malayan Archipelago, especially Malaysia. In Kuching, it is found in Sg. Kuap.

SCORPAENIFORMES
(Scorpionfish and Flatheads)

Fishes of the order Scorpaeniformes are known for their rather ugly appearance and are often referred to as the "mail-cheeked" fishes as the head and body tend to be spiny or have bony plates. The name Scorpaeniformes comes from the Greek 'skorpaina', 'scorpene', a diminutive of the Greek 'skorpios', meaning scorpion, translating as scorpion shape. Scorpaeniformes contains 35 families.

SYNANCEIIDAE
(Stonefish)

Fishes of the family Synanceiidae are found throughout the Indo-Pacific where they are generally marine inhabitants, but some species are found in freshwater. The skin contains glands, which usually appear as warts. The hypodermic-like spines in the dorsal fin inject a neurotoxic poison from venom glands at the base of the spines. The poison is the most deadly of fish venoms, and can be fatal to humans. Stonefish are usually highly camouflaged, blending in with their surroundings.

Leptosynanceia asteroblepa (Richardson, 1844)
Mangrove Stonefish
'Lepu'* or *'Depu'

Description: 23 cm. The skin is a reddish-brown colour with an old-leather-like texture. The head is large and fairly blunt, with a wide, down-turned mouth. The eyes are small and close together on top of the head,

Mangrove Stonefish (*Leptosynanceia asteroblepa*), *'Lepu'*.

giving the fish its nickname 'stargazer'. The head is bony and these bones appear to form a frill-like pattern of triangular spines that extend from beneath the chin to the top of the head. Some of these bones protrude as spines near the gill flaps. The dorsal fin is made up of short distinctive hypodermic-like spines that extend the length of the back to the base of the caudal fin. These spines inject a venom which causes agonizing pain at best or death within one to two hours. The pectoral fin is fan-like, and appears to be used to support the body on the bottom. The anal fin is almost half the length of the dorsal and reaches close to the base of the caudal fin.

Habitat: The muddy environments of coastal and estuarine waters.

Diet: Unknown.

Range: Found throughout the coastal waters of the Indian Ocean, Indonesian Sea, South China Sea and Pacific Ocean. In Kuching, it is found in Sg. Kuap.

Leptosynanceia asteroblepa is well camouflaged for muddy estuarine habitats and produce a distinctive grunting sound when pulled from the water. When disentangling this species from fishing nets great care is needed, although it does not tend to struggle. During handling it has been observed to excrete a milky white fluid from the skin.

PLATYCEPHALIDAE
(Flatheads)

Fishes of the family Platycephalidae are found in marine and estuarine waters throughout the Indo-Pacific. Members of this family have a moderately to strongly depressed head. They are generally demersal, burying themselves in the soft bottom sediments. The name Platycephalidae comes from the Greek, 'platys' meaning flat and 'kephale' meaning head, hence the name flathead. There are 18 genera and 60 species in this family.

Platycephalus indicus (Linnaeus, 1758)
Bartail Flathead
'Baji-baji'

Description: 100 cm. *Platycephalus indicus* is easily identified by the distinctive colour pattern on the caudal fin, which has 2 to 3 horizontal yellow and black stripes on the upper and lower margins. The head is strongly depressed, with two preopercular spines, the lower usually being the longer. These spines, located just before the gills, often become entangled in fishing nets. The pelvic fins are situated behind the pectoral base. There are two dorsal fins, the foremost having 6 to 9 spines with the first spine short and barely connected to the remaining fins. The second dorsal fin has 11 to 15 soft rays. The foremost scale of the lateral line usually has a small spine or ridge. This fish is coloured to blend into a sandy, gravelly bottom. The dorsal region is a brown and tan speckled with transverse darker bands, while the ventral region is pale to white in colour.

Bartail Flathead (*Platycephalus indicus*), *'Baji baji'*.

Habitat: Prefers sandy and muddy bottoms of coastal waters. It regularly enters estuaries, while juveniles have been found in freshwater.

Diet: Unknown.

Range: Found throughout the Indo-West. It has also been introduced into the eastern Mediterranean Sea. In Kuching, it is found in Sg. Kuap.

Platycephalus indicus is commercially cultured in Japan and used in the production of Chinese medicines.

PLEURONECTIFORMES
(Flatfish)

Fishes of the order Pleuronectiformes are almost all marine inhabitants with only four species that may possibly occur in freshwater. They are very distinctive, with a strongly flattened body, where one eye migrates to the other side of the head in adults. The flattened body is somewhat rounded on the side with both eyes, but flat on blind side. The eyes can protrude above the body surface, allowing the fish to see when buried in the substrate. Both the dorsal and anal fins have long bases.

The name Pleuronectiformes comes from the Greek word 'pleura' meaning 'similar, ride, side' and 'nektos' meaning 'to swim', translating as to swim on the same side. There are 11 families in this order.

CYNOGLOSSIDAE
(Tonguefish)

Fishes of the family Cynoglossidae are found throughout tropical and subtropical waters, where they inhabit marine, brackish and freshwater environments. The majority of species occur in shallow coastal waters or estuaries. As the name 'Tonguefish' suggests, these fishes have a tongue-shaped body, with the dorsal and anal fins merging with the caudal. Both the eyes are found on the left side of the head; the eyes are very small and close together. The mouth is asymmetrical and the right pelvic fin is usually undeveloped.

The name Cynoglossidae is derived the Greek word 'kyon' meaning dog, and 'glossa' or tongue, translating as dog tongue. The species found in Kuching could not be identified with confidence as there are many species that are very similar in appearance.

Cynoglossus sp. 1
Tonguefish
'Lidah'

Description: *Cynoglossus* sp. 1 is a flat elongated fish which has been found between 25 and 35 cm. The body is a uniform brown on the dorsal surface and paler on the ventral surface. The snout is pointed, as is the tail. The dorsal and anal fins both have a long fin base and converge with the caudal fin.

Habitat: A benthic species. Thought to be a coastal and estuarine inhabitant.

Diet: Unknown.

Range: This species is relatively common in the estuarine habitats of Sg. Sarawak and Sg. Kuap.

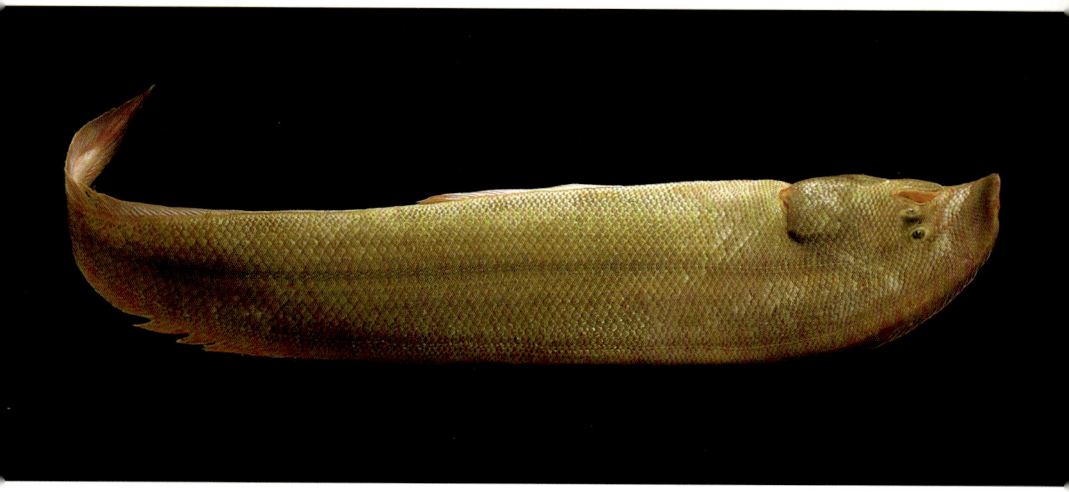

Tonguefish (*Cynoglossus* sp. 1), *'Lidah'*.

Cynoglossus sp. 2
Tonguefish
'Lidah'

Description: This *Cynoglossus* species is slenderer than the one described above, with a pointed snout and tail. The body is also brown, but has darker bands on either side of the lateral line on the dorsal surface.

Habitat: A benthic species. Thought to be a coastal and estuarine inhabitant.

Diet: Unknown.

Range: This species is found occasionally in the estuarine environment of Sg. Sarawak and Sg. Kuap.

Tonguefish (*Cynoglossus* sp. 2), *'Lidah'*.

SOLEIDAE
(Soles)

Fishes of the family Soleidae are found in the tropical and subtropical waters from Europe to Australia and Japan. The body is extremely flattened, with both eyes on the right side of the head. The margin of the pre-operculum is completely concealed (covered by skin and scales). The dorsal fin origin is above or before the eyes. The Soles are benthic dwellers, generally found in the shallower seas of the continental shelf (in waters less than 200 m deep), Their diet consists of benthic invertebrates and fishes. They spawn in pelagic waters

The name Soleidae is derived from the Latin 'solea' meaning sandal, due to the flat shape of this fish. There are 22 genera and 89 species in this family at present, but the taxonomy of this family needs revision.

Soleidae sp. 1
Sole
'Lidah'

Description: This rather unusual specimen has been difficult to identify even to genus, and at approximately 15 cm in length it is likely to be juvenile. The body is teardrop in shape, rounded at the head and pointed at the tail. The dorsal side is a tan, reddish-brown colour, with no obvious markings. The ventral side is a pale pink colour. The dorsal and anal fins almost encircle the entire body apart from the head, both combining with the caudal fin. The lateral line is straight, running mid body.

Habitat: A benthic species, thought to be a marine but an occasional visitor to estuarine waters.

Diet: Unknown.

Range: In Kuching, this species was found in the estuarine waters of Sg. Kuap.

PARALICHTHYIDAE
(Large-tooth Flounders)

Fishes of the family Paralichthyidae are found throughout the Atlantic, Indian and Pacific Oceans. The majority of species are marine inhabitants, rarely entering freshwater. The body is extremely flattened with the eyes on the left side of the head. Their eggs have a single oil globule in the yolk.

The name Paralichthyidae is derived from the Greek 'paralia' meaning seaside or coast and 'ichthys' meaning fish, translating as coastal fish. There are 16 genera and 86 species in this family.

Pseudorhombus arsius (Hamilton, 1822)
Deep Flounder
'Lidah'

Description: 45 cm. The dorsal fin starts near the upper jaw and extends to the caudal peduncle with 71 to 84 rays. The anal fin starts level with the gills and extends to the caudal peduncle with 53 to 62 rays. The caudal peduncle is narrow and the caudal fin is rhomboid.

The dorsal and ventral view of a juvenile sole, *'Lidah'*.

Deep Flounder (*Pseudorhombus arsius*), *'Lidah'*.

HABITAT: Shallow waters of the continental shelf and estuaries. Juveniles are commonly found in brackish waters. It is a benthic inhabitant, dwelling on mud and sandy bottoms.

DIET: Feeds mostly on benthic organisms.

RANGE: Found throughout the Indo-West Pacific. In Kuching, it is found in Sg. Kuap.

TETRAODONTIFORMES
(Pufferfish and Filefish)

Fishes of the order Tetraodontiformes are found both in marine and freshwater environments. There are nine families in this order. The name Tetraodontiformes comes from the Greek 'tetra' meaning four and 'odontos' meaning teeth. This refers to the four teeth, two upper and two lower, that have semi-fused to form a sort of beak, a feature characteristic of this order.

TETRAODONTIDAE
(Pufferfish)

Common and often engaging inhabitants of coral reefs, Pufferfish are chiefly marine, although many species enter and inhabit brackish and freshwater environments. They are found in the tropical and subtropical areas of the Atlantic, Indian and Pacific oceans.

Fishes of the family Tetraodontidae have short, oblong and sub-cylindrical bodies which, in many species, are covered in small spiny bristles. The scales are modified to form plate-like armour over parts of the body. The dorsal and anal fins are situated far back on the body and there are no pelvic fins. The pectoral fin is able to rotate to some degree, and all the fins move in a constant wave-like motion, which enables the puffers to hover, reverse and stop short. As their name suggests, Pufferfish are famous for their ability to 'puff up' to over twice their original size, a defence mechanism designed to deter predators. They inflate their bodies by flooding the folded, concertinaed-like membrane of the swim bladder. Contrary to popular belief, Pufferfish cannot inflate and deflate their bodies without some degree of trauma, and sometimes this may even result in their death.

Many species of Tetraodontidae contain Tetrodotoxin in their flesh, a deadly poison found especially in the viscera, gonads and skin. Pufferfish are a delicacy in Japan, where they are known as 'fugu', and fugu chefs must be specially trained and licensed in the preparation to effectively remove all the poisonous parts of the fish. Many deaths are reported annually from the consumption of these fishes. In Malaysia Pufferfish are known as '*Ikan Buntal*' and are not eaten except for *Lagocephalus scleratus*, which is consumed in Sarawak.

The Tetraodontidae are omnivorous and, depending on the habitat, will eat just about anything such as algae, molluscs, crustaceans, worms and sponges. They are also opportunistic scavengers, and many are attracted to dying fish stuck in nets. They lay their eggs in protected niches of the substrate, where one or both parents are thought to guard them.

Tetraodon nigroviridis (Marion de Procé, 1822)
Spotted-green Pufferfish
'Buntal bako'

DESCRIPTION: 90 cm. Commonly caught at 20 to 30 cm in length. The head is rounded with large eyes, and the nasal organs have distinctive papillae. The body is rounded and covered in large and small black or dark brown spots encircled by yellow rings. In juveniles, this pattern is very clear, whereas in adults the sides of the body tend to become a dirty brown colour, varying with individuals. The belly is an off-white colour.

Spotted-green Pufferfish (*Tetraodon nigroviridis*), *'Buntal bako'*.

Habitat: Coastal and estuarine waters. large groups of juveniles are often found upstream, in shallow rivers.

Diet: Omnivorous.

Range: Found throughout Asia. In Kuching, it is found in Sg. Kuap and the more brackish waters of Sg. Sarawak.

The adults of *Tetraodon nigroviridis* are often caught in nets attracted by the struggling and dying fish that have already been caught. When they are pulled out of the water in nets, they produce a peculiar creaking noise as air is inhaled and exhaled.

Xenopterus naritus (Richardson, 1848)
Yellow Pufferfish
'Buntal pisang'

Description: 28 cm. The body is more 'torpedo'-shaped than some species, with a bright golden colour, darker above and lighter below. The dorsal and anal fins are almost symmetrical in size and position, while the

Yellow Pufferfish (*Xenopterus naritus*), *'Buntal pisang'*.

anal fin is just slightly shorter and smaller. The nasal organs are large circular tyre-like disks. The mouth is terminal and blunt.

Habitat: Coastal and estuarine species, found in the lower reaches of tidally influenced rivers.

Diet: Omnivorous.

Range: In Kuching, it is found in Sg. Kuap and the more brackish waters of Sg. Sarawak.

Xenopterus naritus is often caught in nets where already entangled fish offer a free meal.

Lagocephalus lunaris (Bloch & Schneider, 1801)
Green Rough-backed Pufferfish
'Buntal'

Description: 45 cm, although generally caught between 3 and 20 cm. The body is relatively elongate with a broad head which is slightly box-shaped, giving it a rather bulldog attitude. The eyes are large and rounded with a free orbital margin on the lower side. The nasal organ in front of the eye consists of a short papilla with two large openings.

The dorsal region from the interorbital space behind the nostrils to the origin of the dorsal fin is covered with prickles. Similarly, the ventral region

Green Rough-backed Pufferfish (*Lagocephalus lunaris*), *'Buntal'*.

from below the anterior margin of eye to the anterior of the anus is also covered. There are two lateral lines: the upper is conspicuous, extending from the top of the head and curves down to the flank where it reaches the caudal peduncle; the lower lateral line is indistinct. The dorsal and anal fins are short and set far back on the body. The caudal peduncle is compressed and the caudal fin slightly forked.

This species is attractive, with a metallic gold colour, darker on the back and sides to an almost silvery gold colour underneath. A wide silvery band runs longitudinally along the mid-lateral line from the mouth to the base of the caudal fin. The dorsal fin has 11 to 13 soft rays while the anal fin has 10 to 12 soft rays. *Lagocephalus lunaris* can be differentiated from *L. enermis* by the presence of prickles on the back extending from nape to the caudal peduncle, while the back of *L. enermis* is smooth.

Lagocephalus lunaris in all its puffed-up glory; when it inflates its swim bladder the small stubble-like prickles along the belly are clear.

Habitat: Mainly marine. The young regularly enter estuaries. Juveniles are commonly caught in the cast nets (*jala*) in the upper regions of tidally influenced rivers.

Diet: Omnivorous.

Range: Found throughout the Indo-Pacific Ocean and is common around Peninsular Malaysia and Sarawak. In Kuching, it is found in Sg. Kuap and the more brackish waters of Sg. Sarawak.

Carinotetraodon borneensis (Regan, 1903)

Redeye Pufferfish

'Buntal'

Description: 6 cm. It has a bullet-shaped body. The dorsal fin has 12 to 13 soft rays, the anal fin has 10 to 12 soft rays and the pectoral fin has a black base and 15 rays. The male has red dorsal and anal fins, while the caudal fin is dark grey with a white margin. The individuals caught around Kuching all have brown bodies with a faint or strong yellow triangle pattern on the head.

Redeye Pufferfish (*Carinotetraodon borneensis*), *'Buntal'*.

Some individuals have bright red bellies, while many have been observed to attain this colouring as they grow in captivity. The eyes appear blind, with an opal-like blue colour and a circumferential red ring.

Habitat: Freshwater; small to large sized rivers extending down to upper tidal reaches.

Diet: Feeds on zooplankton, molluscs, crustaceans and other invertebrates.

Range: Is thought to be found throughout Asia, but currently has been confirmed only from south-western Sarawak. In Kuching, it is found in the fresher waters of Sg. Sarawak.

Carinotetraodon borneensis, a popular species in the aquarium trade, used to be known as *C. lorteti* (Redeye Pufferfish) but recently *C. lorteti* has been restricted to the species found in China. These species are fairly difficult to differentiate as there is a marked morphological difference between males and females, plus the fact that they can change colour in order to blend with their surroundings. They also appear to enhance their colours as they become dominant either sexually or territorially. *C. salivator* is also found in the outskirts of Kuching but in shallower and clearer waters than the rivers running through Kuching proper.

It is found in large groups as juveniles, but singly as adults. Aggressive behaviour has been observed among adults. Two contesting individuals may square off to one another, folding their skin laterally along both the belly and the dorsal region to form what appear to be additional fins (a yellow fin dorsally and a red fin ventrally). This behaviour lasts until one individual gives ground and is chased away. The dominant individual is usually bigger and has the brighter colour pattern: the yellow triangle being more pronounced, a brighter red belly and a stronger white margin to the tail.

BATRACHOIDIFORMES

Fishes of the order Batrachoidiformes are found in mostly marine and brackish environments although there are a few totally freshwater species in South America. There is only one family in this order, and the name Batrachoididae is derived from the Greek 'batrachos' meaning frog, literally translated 'frog shape'. This name was most likely given for the loud croaking sound these fish make when disturbed.

BATRACHOIDIDAE
(Toadfish)

Members of the Toadfish family are small- to medium-sized fish, with broad, flattened heads, often covered in barbels or fleshy appendages. Sharp spines are found around the opercle and subopercle, which are venomous only in members of the subfamily Thalassophryninae ("venomous toadfishes"). They are mostly drab brown or black in colour with variable markings such as blotches, saddles and bars, with the exception of some brightly coloured species in the Caribbean. A benthic fish, Toadfish are well camouflaged for rocky bottoms where they wait to ambush small fish, molluscs and crustaceans.

The males of some species of Toadfish, if not all, make a loud droning sound or 'singing' to attract females during courtship. The call, especially at night, has been reported to be so loud in some species as to be annoying to nearby shore dwellers and houseboat inhabitants. This sound is produced by the high frequency vibrations (6000 beats per minute, i.e., twice that of the hummingbird's wings) of the muscles around the swim bladder, and these amazing muscles can keep vibrating for periods of over an hour at a time. It is hoped that knowledge of how these muscles operate may one day be used to help in finding possible cures for muscle-related diseases in humans.

Halophryne hutchinsi (Greenfield, 1998)
Grunting Toadfish
'Bengut'* or *'Gedempu'* or *'Gedema'?

Description: 30 cm. The colour is an irregular patterning of light and dark browns, with lighter saddles and blotches on the back and body. The fleshy hair-like growths on the head may be sensitive to water movement, and hence help to detect prey. The dorsal and anal fins have alternating light and dark bands, while the pectoral, pelvic and caudal fins have dark stripes.

(Opposite). Grunting Toadfish (*Halophryne hutchinsi*), *'Bengut'*. The close-up of the head shows the fleshy hair-like growth that may be sensitive to water movement and hence help detect prey.

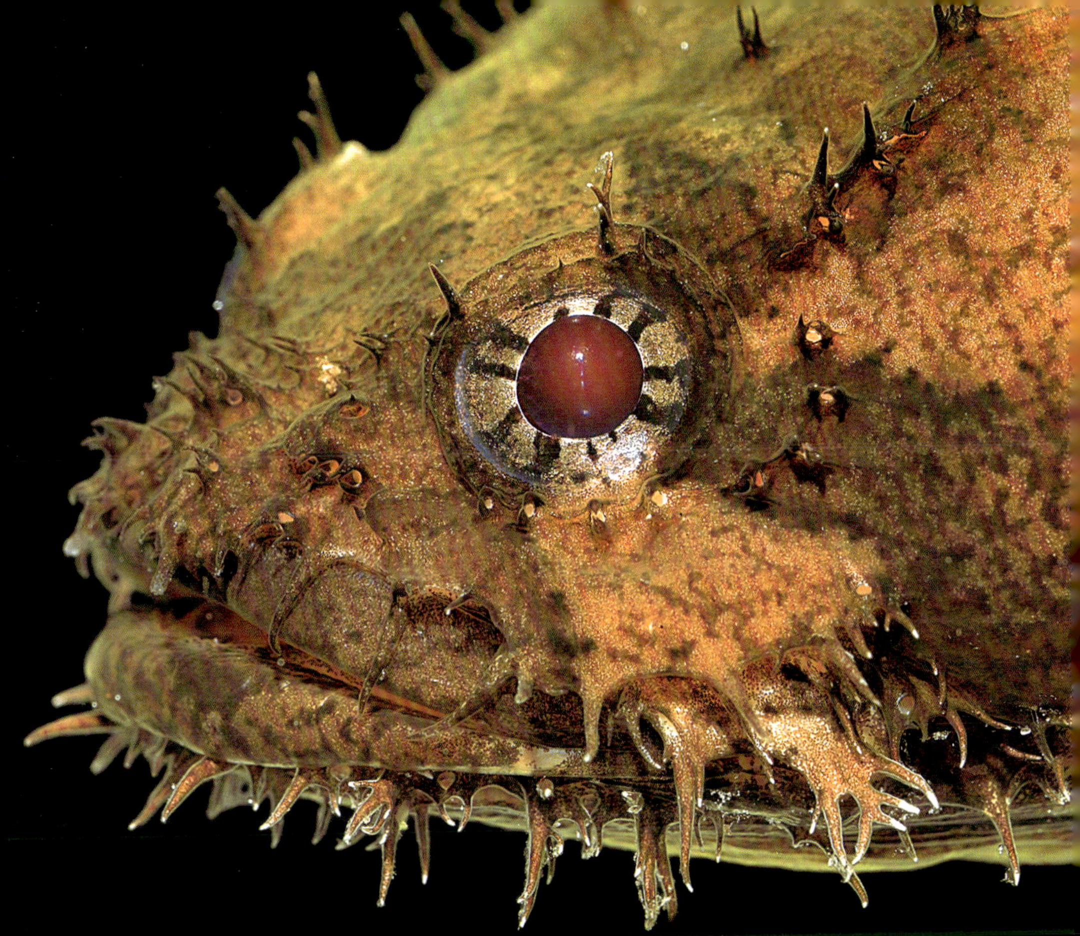

The pectoral fin has 19 to 21 rays, the anal fin has 14 to 15 rays, the second dorsal fin has 17 to 19 rays and the lower lateral line has 22 to 25 pores.

Habitat: Marine and estuarine.

Diet: Unknown.

Range: Found in the estuarine waters of the Kuap River, this is the first recording for Sarawak and Malaysia; previously it was only known from the Philippines and Pulau Waigeo (Indonesia). *Allenbatrachus grunniens* is also found throughout Malaysian waters, and *Batrachomoeus trispinosus* may also be present.

This species is fairly new to the taxonomic world, first described from the Philippines and Indonesia by Dr D. Greenfield in 1998.

Halophryne diemensis is the most similar species to *Halophryne hutchinsi*, but has not been confirmed in this area. *H. diemensis* has more rays in its fins and more pores in its lateral line: 22 or more in the pectoral fin, 16 or more in the anal fin, 19 or more in the dorsal fin and 33 to 37 pores in the lower lateral line.

Acknowledgements

Firstly, let me express my gratitude to YB Datuk Michael Manyin Ak Jawong, Minister for Environment and Public Health, the Sarawak Government, the Natural Resources and Environment Board Sarawak (NREB) and the Danish International Development Assistance (DANIDA) for their foresight in initiating and supporting this project, and for assigning Chemsain Konsultant Sdn. Bhd. and myself for the task. I would also like to give a special thank you to Prof. Dr Robert Inger, who has been kind enough to write the Preface.

I would like to express my sincere thanks to the many people who have helped me in identifying the fish and who have used their precious time to read and adjust the descriptions to ensure they are as accurate as possible: Dr Charles Leh of the Sarawak Museum and Dr Kelvin Lim of the University of Singapore for their amazing patience and encouragement over the many visits and emails. Dr Helen Larson, Dr Patricia Kailola, Dr David Woodland, Dr David Greenfield, Dr John Randal, Dr Bruce Collette, Dr Ian Harrison, Dr Sven Kullander, Dr John McEachran, Dr Tan Heok Hui, and Dr Ng Peter Kee Lin for their knowledge and willingness to help.

I would also like to thank Dr Rachel Atanacio of FishBase and the FishBase website team. Their website has been an endless source of information on the fish, and their assistance to find help from specialised fish taxonomists whenever needed has been indispensable.

I would like to thank all those involved in collecting the specimens and who helped identify the local names for each species, especially Hj. Wailei bin Hj. Bujang and Mohammad Zen Ben Omar. I would also like to thank them for their skills as fishermen and boatmen on Sg. Sarawak and Sg. Maong; I would like to thank Captain Goh for his help and support in providing very able fishermen and boatmen for Sg. Kuap; and the technicians from Chemsain who have helped with fishing and translation in the field, especially Marcus Gajis.

For the most outstanding photography that has really made this book special, thank you Ch'ien C. Lee, amazing! Also I would like to thank Jason Isley and Roger Munns from Scubazoo for the marine photographs on pages 18–20.

ACKNOWLEDGEMENTS

Special thanks to Erling Poulsen from COWI Denmark for years of communication and support regarding the environmental quality of the Sarawak River and its importance for the fish fauna.

Also special thanks to my linguistic, editorial and emotional support group: Warwick Anderson, Fairul Reeza bin Mohd Rosli, Rosanna Chio, Madeleine Yong, Glen MacNair, Jakob Larsen and Jude Bong and last, but not least my family.

And finally, I would like to thank Ib Larsen, who gave me the opportunity, encouragement and support to write this book, and Datuk C.L. Chan, my publisher, and his team and Kay Lyons, a very thorough and excellent editor, for making this book into this wonderful creation you see before you.

THANK YOU!

Glossary

Abdominal: on or pertaining to the belly or abdomen.
Adult: sexually mature individual.
Anadromous: fish that migrate from the sea to freshwater in order to spawn.
Anterior: on or towards the front (head) end of the fish. Opposite of posterior.
Barbel: slender tentacle-like protuberance or fleshy projection near mouth, chin or snout.
Benthic: dwelling on, or relating to, the bottom of a body of water.
Brackish water: a habitat characterised by mixing of fresh and salt water.
Carnivore: feeds on other animals.
Catadromous: fish that migrate from freshwater to the sea in order to spawn.
Chromatophore: coloured or pigmented cell in the skin.
Compressed: flattened laterally (from both sides).
Cutaneous: pertaining to the skin (dorsal can also be used).
Demersal: living/lying near or on the bottom.
Depressed: flattened ventrally (from top to bottom).
Diurnal: active during the day.
Dorsal: towards the back or upper part of the body.
Endemic: native and restricted to a particular area.
Filament: a thread-like, slender or elongated appendage.
Fins: see page 24.
Fontanelle: unossified median opening on the top of the skull.
Gill rakers: bony, finger-like projections of the gill arch. The number and length of these are important in the classification and identification of fishes.
Herbivore: feeds on plant material.
Hermaphrodite: to have both sexes in the same individual, either at the same time (synchronous hermaphrodite) or at different times (successive hermaphrodite).
Interorbital: the space between the eyes.
Isthmus: the fleshy throat region of a fish.

Juvenile: a sexually immature individual.

Lachrymal: a dermal, lateral-line bone situated directly in front of the eye socket or 'orbit'. Also called preorbital, lachrimal, and lacrimal.

Lateral: refers to the side or directed toward the side or extending from the side; the opposite of medial.

Mandible: the main bone in the lower jaw; 'mandibulary', pertaining to the lower jaw.

Margin: the edge or rim.

Maxillary: the lateral part of the upper jaw.

Medial: toward the middle or median plane of the body; opposite of lateral.

Mental: relating to the chin.

Nasal: referring to the nostrils or the olfactory organ.

Nictitating eyelid: a transparent inner eyelid.

Nocturnal: active during the night.

Oesophageal: pertaining to the oesophagus (section of alimentary canal between the mouth and the stomach).

Omnivore: feeding on both plants and animals.

Operculum: bony gill cover, comprising four bones: opercle, preopercle, interopercle and subopercle.

Ovo-viviparity: internal fertilisation, where the female keeps the eggs inside until they hatch.

Palantine: relating to the bone on the roof of the mouth lying between the vomer and the upper jaw.

Papilla: small, nipple-like, fleshy projection (plural: papillae).

Pelagic: referring to the surface or middle depths of a body of water.

Pharyngeal: relating to or found in the region of the pharynx.

Piscivore: prefers to feed on fish.

Planktivore: feeding on plankton.

Posterior: on or towards the tail end of the fish.

Predominate: most important, most dominant.

Rugose: having a rough or wrinkled surface.

Scute: hard, thick external plate on the skin. In fish it is often an enlarged, modified, thickened scale that can be keeled or spiny; usually associated with the lateral line, or at the side in the tail region, or along the ventral profile.

Serrated: having a saw-toothed edge.

Spinule: a small spine.

Spiracle: a respiratory opening situated between the eye and the first gill slit, usually associated with sharks and rays.

Standard length: measurement from the snout to the base of the tail.

Terminal: situated at or forming the end.

Tubercle: wart-like protrusion or bump in the skin.

Ventral: referring to the lower part/lower surface/underside of the body.

Viviparous: bringing forth living young, rather than laying eggs.

Vomer: a median unpaired bone on the roof of the mouth; its anterior end often bears teeth.

Selected References

Allen, G. (2000). *Marine Fishes of South-East Asia.* Periplus Editions (HK) Ltd., Singapore.

Atack, K. (2005). *Monitoring Abundance of Fish, Shrimps and Prawns in Rivers and Tributaries in Kuching* (October, 2003–August, 2004). Natural Resources and Environmental Board (Sarawak). In press.

Carpenter, K.E. and Niem, V.H. (eds.) (1999 & 2001). FOA Species Identification Guide for Fishery Purposes, The Living Marine Resources of the Western Central Pacific, Vol. 3 & 4 (1999) and 5 & 6 (2001).

Chapman, P. (1985). Cave-frequenting vertebrates in the Gunung Mulu National Park, Sarawak. *Sar. Mus. Journ.* 34(55): 101–113.

Chew, S.F. and Ip, Y.K. (1992). Biochemical adaptations of the mudskipper, *Boleophthalmus boddarti*, to a lack of oxygen. *Marine Biology* 112: 567–571.

Collette, B.B. (2004). Family Hemiramphidae Gill 1859. Annotated Checklists of Fishes. No 22. California Academy of Sciences. http://www.calacademy.org/research/ichthyology/annotated/

Cramphorn, J. (1978). A preliminary survey of the fishes in Gunung Mulu National Park, Sarawak. 1977/78. (unpublished report)

Dipper, F. (2001). *Extraordinary Fish.* BBC Worldwide Ltd.

Doi, A., Iwata, T., Inoue, M., Miyasaka, H., Sabki, M.S. and Nakano, S. (2001). A collection of freshwater fishes from the Rayu Basin of Western Sarawak, Malaysia. *Raffles Bull. Zool.* 49(1): 13–17.

Froese, R. and Pauly, D. (eds.). 2005. FishBase. World Wide Web electronic publication. www.fishbase.org, version (02/2005).

Greenfield, D.W. (1998). *Halophryne hutchinsi*: A New Toadfish (Batrachoididae) from the Philippine Islands and Pulau Waigeo, Indonesia. *Copeia* 3: 696–701.

Gremli, M.S. and Nerman, H.E. (2001). *Marine Life of the South China Sea.* Insight Print Services (Pte.) Ltd. or APA Publications GmbH & Co., Verlag KG.

Grinang, Jongkat and Lim, K.K.P. (2004). Fishes in: Sarawak Bau Limestone Biodiversity. *Sar. Mus. Journ.* Vol 59, 80 (New Series), Special Issue No. 6. Sarawak Biodiversity Centre, Kuching.

Hazebroek, H.P. and Abang Kashim bin Abang Morshidi (2001). *National Parks of Sarawak.* Natural History Publications (Borneo), Kota Kinabalu.

Ho, A.Y. (2004). *Old Kuching.* 2nd Edition. Oxford University Press, New York.

Inger, R.F. and Chin, P.K. (1962). *The Freshwater Fishes of North Borneo.* Field Museum of Natural History, Chicago. Reprinted 2002 by Natural History Publications (Borneo), Kota Kinabalu.

Kottelat, M. and Lim, K.K.P. (1995). Freshwater fishes of Sarawak and Brunei Darussalam: A preliminary annotated checklist. *Sar. Mus. Journ.* 48(69): 227–256.

Kottelat, M., Whitten, A.J., Kartasari, S.N. and Wirjoatmodjo, S. (1993). *Freshwater Fishes of Western Indonesia and Sulawesi.* Periplus Editions (HK) Ltd., Singapore.

Larson, H.K. and Lim, K.K.P. (2005). *A Guide to the Gobies of Singapore.* Singapore Science Centre, Singapore.

Lee Nyanti, Ling Teck Yee and Khairul Adha (1999). Freshwater fishes from Bario, Kelabit Highlands Sarawak. *Asean Review of Biodiversity and Environmental Conservation.* http://www.arbec.com.my/pdf/art4sepoct99.pdf

Lim, K.K.P. and Ng, P.K.L. (1990). *A Guide to the Freshwater Fishes of Singapore.* Singapore Science Centre, Singapore.

MacKinnon, K., Hatta, G., Halim, H. and Mangalik, A. (1996). *The Ecology of Kalimantan.* Periplus Editions (HK) Ltd., Singapore.

Meisner, A.D. and Collette, B.B. (1998). A New Species of Viviparous Halfbeak, *Dermogenys Bispina* (Teleostei: Hemiramphidae) from Sabah (North Borneo). *Raffl. Bull. Zool.* 46(2): 373–380.

Mohsin, Abu Khair Mohammad and Ambak, Mohd. Azmi (1996). *Marine Fishes and Fisheries of Malaysia and Neighbouring Countries.* Universiti Pertanian Malaysia Press, Serdang.

Roberts, T.R. (1989). *The Freshwater Fishes of Western Borneo (Kalimantan Barat, Indonesia).* California Academy of Science, San Francisco.

Scott, J.S. (1959). *An Introduction to the Sea Fishes of Malaya.* Government Press, Kuala Lumpur.

Watson, D.J. and Balon, E.K. (1984). Structure and production of fish communities in tropical rain forest streams in Northern Borneo. *Can. J. Zool.* 62: 927–940.

Woodland, D.J., Cabanban, A.S., Taylor, V.M. and Taylor, R.J. (2002). A synchronised rythmic flashing light display by schooling *Leiognathus splendens* (Leiognathidae: Perciformes). *Mar. Freshwater Res.* 53: 159–162.

Index

Titles by Natural History Publications (Borneo)

For more information, please contact us at

Natural History Publications (Borneo) Sdn. Bhd.
A913, 9th Floor, Phase 1, Wisma Merdeka
P.O. Box 15566, 88864 Kota Kinabalu, Sabah, Malaysia
Tel: 088-233098 Fax: 088-240768 e-mail: info@nhpborneo.com

www.nhpborneo.com

SABAH COLOUR GUIDE
KUDAT
Wendy Hutton
Natural History Publications (Borneo)

沙巴景区彩色导游
古达

SABAH COLOUR GUIDE
SANDAKAN
History, Culture, Wildlife and Resorts of the Sandakan Peninsula
Wendy Hutton

沙巴景区彩色导游
山打根

National Parks of SARAWAK

GUNUNG MULU NATIONAL PARK

MALIAU BASIN
SABAH'S LOST WORLD

A Walk through the LOWLAND RAIN FOREST of Sabah
Elaine J.F. Campbell

PREFERRED CHECK-LIST OF SABAH TREES
Third Edition
Y.F. LEE
Natural History Publications (Borneo)

IN BRUNEI FORESTS
An Introduction to the Plant Life of Brunei Darussalam
K.M. Wong
with watercolours by C.L. Chan
A Revised Edition

THE LARGER FUNGI OF BORNEO
David N. Pegler

RAFFLESIA OF THE WORLD
JAMILI NAIS

PITCHER-PLANTS OF BORNEO
Anthea Phillipps and Anthony Lamb

NEPENTHES of BORNEO
CHARLES CLARKE
Natural History Publications

A GUIDE TO THE
Pitcher Plants of Sabah
Charles Clarke
Natural History Publications (Borneo)

A GUIDE TO THE
Pitcher Plants of Peninsular Malaysia
Natural History Publications (Borneo)

NEPENTHES
OF SUMATRA
AND PENINSULAR MALAYSIA
Natural History Publications (Borneo)

BORNEO
HUGO STEINER
ITS MOUNTAINS AND LOWLANDS
WITH THEIR PITCHER PLANTS
TREKKING FROM 1962 TO 2007

ORCHIDS
of SARAWAK
TEOFILA E. BEAMAN, JEFFREY J. WOOD
REED S. BEAMAN AND JOHN H. BEAMAN
Natural History Publications (Borneo)
The Royal Botanic Gardens, Kew

ORCHIDS
of SUMATRA
J.B. Comber

DENDROCHILUM
OF BORNEO
Jeffrey J. Wood
Natural History Publications (Borneo)
in association with
The Royal Botanic Gardens, Kew

THE PLANTS of
MOUNT KINABALU
3. GYMNOSPERMS AND NON-ORCHID MONOCOTYLEDONS
John H. Beaman and Reed S. Beaman
Natural History Publications (Borneo)
in association with
The Royal Botanic Gardens, Kew

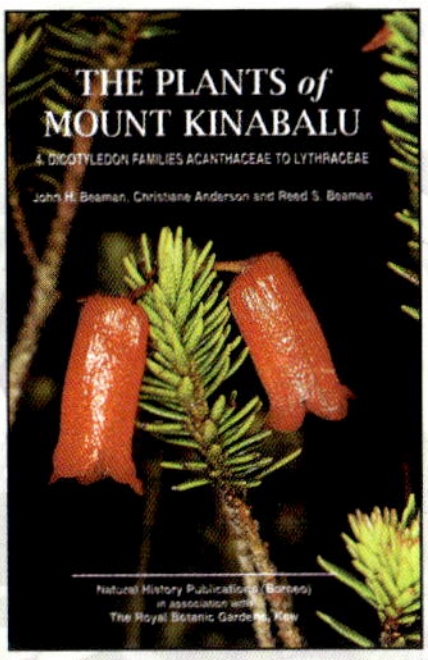
THE PLANTS of
MOUNT KINABALU
4. DICOTYLEDON FAMILIES ACANTHACEAE TO LYTHRACEAE
John H. Beaman, Christiane Anderson and Reed S. Beaman
Natural History Publications (Borneo)
in association with
The Royal Botanic Gardens, Kew

The Genus
PAPHIOPEDILUM
Second Edition
Phillip Cribb
Natural History Publications (Borneo)
The Royal Botanic Gardens, Kew

SLIPPER ORCHIDS
OF BORNEO
Phillip Cribb
Natural History Publications

COELOGYNE
A SYNOPSIS
Dudley Clayton
Natural History Publications (Borneo)
in association with
The Royal Botanic Gardens, Kew

ORCHID
CONSERVATION
Editors
Natural History Publications (Borneo)

GINGERS
of PENINSULAR MALAYSIA
AND SINGAPORE
K. Larsen, H. Ibrahim, S.H. Khaw and L.G. Saw
Natural History Publications (Borneo)

MOSSES AND LIVERWORTS
of MOUNT KINABALU
Jan Peter Frahm, Wolfgang Frey,
Harald Kürschner and Mario Menzel

PHASMIDS
of BORNEO
Philip E. Bragg
Natural History Publications (Borneo)

An Illustrated Guide to the
STICK AND LEAF INSECTS
of Peninsular Malaysia and Singapore
FRANCIS SEOW-CHOEN
Natural History Publications (Borneo)

A GUIDE TO THE
DRAGONFLIES
OF BORNEO
THEIR IDENTIFICATION AND BIOLOGY
A.G. ORR
Natural History Publications (Borneo)

A Guide to Australian
GRASSHOPPERS AND LOCUSTS
D. C. F. Rentz, R. C. Lewis, Y. N. Su and M. S. Upton
Natural History Publications (Borneo)

The
BIRDS of BORNEO
Natural History Publications (Borneo)
in association with The Sabah Society

The
BIRDS of BURMA
Fourth Edition
Bertram E. Smythies
Natural History Publications (Borneo)

Swiftlets of Borneo
BUILDERS OF EDIBLE NESTS
Lim Chan Koon and Earl of Cranbrook
Natural History Publications (Borneo)

The Natural History of
ORANG-UTAN
Elizabeth L. Bennett
Natural History Publications (Borneo)

PROBOSCIS MONKEYS OF BORNEO
Elizabeth L. Bennett and Francis Gombek

A Field Guide to the
FROGS OF BORNEO
Robert F. Inger
Robert B. Stuebing

Panduan Lapangan
KATAK-KATAK BORNEO
Robert F. Inger
Robert B. Stuebing
Disunting oleh
Anna Wong
Shahrul Anuar Mohd. Sah
Natural History Publications (Borneo)
Jabatan Muzium Sabah

The Natural History of
AMPHIBIANS AND REPTILES
IN SABAH
Robert F. Inger and Tan Fui Lian

An Introduction to the
Amphibians and Reptiles
of Tropical Asia
Indraneil Das
Natural History Publications (Borneo)

A FIELD GUIDE TO THE
SNAKES of BORNEO
Robert B. Stuebing and Robert F. Inger
Natural History Publications (Borneo)

Man-eating
CROCODILES
of Borneo
JAMES RITCHIE
with JOHNSON JONG
Natural History Publications (Borneo)

TURTLES of BORNEO
AND PENINSULAR MALAYSIA
Lim Boo Liat and Indraneil Das
Natural History Publications (Borneo)

A POCKET GUIDE
LIZARDS OF BORNEO
Indraneil Das
Natural History Publications (Borneo)

A POCKET GUIDE
PITCHER PLANTS OF SARAWAK
Charles Clarke and Ch'ien Lee
Natural History Publications (Borneo)

MARINE FOOD FISHES
AND FISHERIES
OF SABAH
Chin Phui Kong
Natural History Publications

THE
FRESH-WATER FISHES
OF NORTH BORNEO
ROBERT F. INGER AND CHIN PHUI KONG
With a Supplementary Chapter by Chin Phui Kong
Natural History Publications (Borneo)

LAYANG LAYANG
A Drop in the Ocean
Nicolas Pilcher, Steve Oakley and Ghazally Ismail
Natural History Publications (Borneo)

Three
Came Home
Agnes Newton Keith
Natural History Publications (Borneo)

LAND BELOW
THE WIND
Agnes Newton Keith
Natural History Publications (Borneo)

WHITE MAN
RETURNS
Agnes Newton Keith
With an Introduction by Patricia Regis
Natural History Publications (Borneo)

WITH THE WILD MEN OF
BORNEO
Elizabeth Mershon
Natural History Publications (Borneo)

FOREST LIFE AND ADVENTURES
IN THE
MALAY ARCHIPELAGO
Eric Mjöberg
Natural History Publications (Borneo)

A NATURALIST
IN BORNEO
Robert W.C. Shelford
Natural History Publications (Borneo)

TWENTY YEARS
IN BORNEO
Charles Bruce
Natural History Publications (Borneo)

The
DRAGON
of Kinabalu
and other Borneo Stories
Owen Rutter
Natural History Publications (Borneo)

A CULTURAL HERITAGE OF NORTH BORNEO
Animal Tales
of Sabah
P. S. Shim
with illustrations by Yong Ket Hyun
Natural History Publications (Borneo)

Kadazan
Folklore
Compiled and edited by
Rita Lasimbang
Illustrated by
Suzie Majikol
Natural History Publications (Borneo)

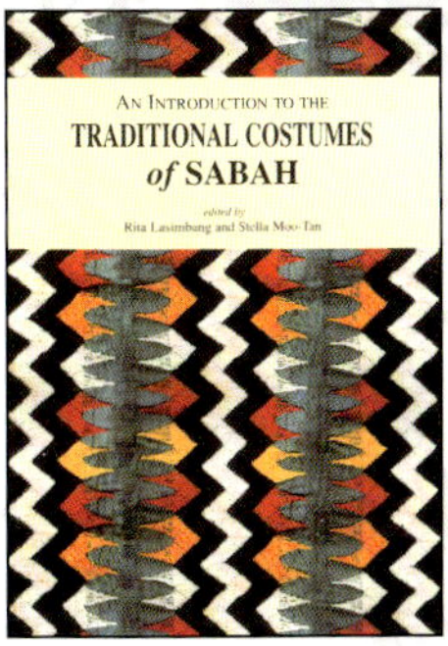
AN INTRODUCTION TO THE
TRADITIONAL COSTUMES
of SABAH
edited by
Rita Lasimbang and Stella Moo-Tan

Etnobotani
GARY J. MARTIN
Diterjemah oleh Maryati Mohamed
SEBUAH MANUAL PEMULIHARAAN
'MANUSIA DAN TUMBUHAN'
Natural History Publications (Borneo)

Water Land Cities
Sabah, Malaysian Borneo
Natural History Publications (Borneo)
Environment Protection Department